10 × 10

Hot Water From the Sun

A Consumer Guide to Solar Water Heating

by
The Franklin Research Center
Philadelphia, PA

for
Division of Energy, Building Technology and
Standards; Office of Policy Development and
Research; U.S. Department of Housing and Urban
Development in cooperation with the U.S.
Department of Energy

Author: Beth McPherson
Illustrator: Dava Lurie

University Press of the Pacific
Honolulu, Hawaii

Hot Water from the Sun:
A Consumer Guide to Solar Water Heating

by
The Franklin Research Center
U.S. Department of Housing and Urban Development
U.S. Department of Energy

ISBN: 1-4102-2037-0

Copyright © 2005 by University Press of the Pacific

Reprinted from the 1980 edition

University Press of the Pacific
Honolulu, Hawaii
http://www.universitypressofthepacific.com

DEPARTMENT OF HOUSING AND URBAN DEVELOPMENT
WASHINGTON, D.C. 20410

ASSISTANT SECRETARY FOR
POLICY DEVELOPMENT AND RESEARCH

TO THE READER:

This guide has been written by experts in the field to provide you, a consumer, with information about the use of solar energy to heat the water you use in your home. It is no secret that energy costs have been rising rapidly in the past several years, and that conventional sources of energy are in short supply.

As more and more people feel the impact of the rising cost and limited availability of some types of energy, increasing attention is being given to the energy available in sunlight. No longer just a fad for some individual experimenters, no longer only a subject for scientific and engineering studies, solar energy applications are now a matter of intense interest to almost everyone.

Of all of the ways in which we can capture and use solar energy to meet our needs, providing hot water is perhaps the simplest and most economical for the homeowner. Many manufacturers now make equipment and systems to provide solar-heated water, and many firms around the country are qualified to install, maintain, and service them. The Federal government and several states even provide tax credits to help cover the cost of solar energy systems.

But deciding to buy and install a solar hot water system still requires you to consider a number of factors—whether the system will save *you* money now or in the future, what kind of system to buy and from whom to buy it, whether to install it yourself or have it installed, and similar matters. This guide is designed to provide some of the information you will need to make these decisions, and to tell you where to get the rest of the information you may need to install or have installed a satisfactory solar hot water system.

I know that you share our concern about energy and that you are looking for ways to save energy and money at the same time. I appreciate your interest in solar energy and I trust that this information will help you make a wise decision.

Donna E. Shalala
Assistant Secretary

Acknowledgments

Grateful acknowledgment is made to Booz-Allen & Hamilton, Inc., which as part of its contract to develop RSVP, the U.S. Department of Housing and Urban Development's computer program for estimating solar economics, produced both the method and the tables used in the economic worksheets in this guide.

Invaluable information and suggestions also came from: Ken Bossong, Citizens' Energy Project; Donald Carr, National Association of Home Builders; Martha Cohen, Connecticut Citizen Action Group; Lynda Connor, Department of Energy; Pat Coughlan, Department of Energy; Joe Dawson, Grumman Energy Systems, Inc.; Randy Dyer, Solar Energy Industries Association; Paul Erhartic, Northeast Solar Energy Center; Martin Giuffrida, Massachusetts Solar Action Office; J. Thomas Greene, House Commerce Subcommittee on Oversight and Investigations; Cedric Grgurich, Solar Energy Industries Association; James Hill, National Bureau of Standards; Paul Kando, PRC Energy Analysis Co.; Molly Kuntz, PRC Energy Analysis Co.; Susannah Lawrence, Solar Lobby; Buddy Mawyer, Sheet Metal and Air Conditioning Contractors' National Association; Kenneth G. Moore, Department of Energy; Edward Moran, Federal Trade Commission; Harold Olin, U.S. League of Savings Associations; Linda O'Neil, Consumer Information Center; Wayne Parker, SolarCal; Kathryn Ramsay, California Department of Consumer Affairs; Lee Richardson, U.S. Office of Consumer Affairs; Heidi Sanchez, Federal Trade Commission; Rick Schwolsky, National Association of Solar Contractors; Robert O. Smith, P.E.; Peter Thorne, Massachusetts Solar Action Office; Rebecca Vories, Solar Energy Research Institute; Preston Welch, Rho Sigma; John Wellinghoff, Federal Trade Commission; Marvin Yarosh, Florida Solar Energy Center.

Hot Water From the Sun

Part Three Protecting Your Investment

Introduction

This guide is addressed to anyone considering the use of a solar energy system to heat the water used in a home—either for a new home or to be installed (or "retrofitted") in an existing home. It has been arranged to give you a general understanding of solar energy and of the various types of systems used to provide solar heated water, to provide you a method to estimate the costs and potential savings from using such systems, and to help you select a specific system and choose the best way to have it installed.

The guide stresses the importance of proper installation, and recommends that you choose a supplier or installation contractor who can be relied on to select a system well suited to your needs. Even so, we believe that you must know the basics of solar water heating, if only to understand the information provided by your contractor and to operate and maintain the system once installed.

Some technical information on system efficiencies has been included as an appendix, and a separate HUD report* provides detailed information on installation methods for the tradesman and the skilled homeowner who plans to install his or her own system.

*Installation Guidelines for Solar DHW Systems in One- and Two-Family Dwellings. 111 pages. Superintendent of Documents, U.S. Government Printing Office, Washington, D.C. 20402, Stock No. 023-000-00520-4: $4.00.

How to Use This Guide

In reading this guide, you should be looking for answers to a series of questions. The principal ones are: Should I install a solar hot water system? How do I get a good system? How can I be sure it will operate effectively? The guide addresses each of these questions in turn.

Part One, "Is Solar Water Heating Right for You?" provides information to help you decide whether to install a solar hot water system.

Chapter 1: How Solar Water Heaters Work

This chapter discusses the components which make up a solar hot water system, the various kinds of systems available, and the benefits and limitations of each type.

Chapter 2: Making Good Use of the Sun

A solar energy system needs sunlight to operate. A house set in a grove of trees which provide constant shade, or arranged on the property so that no generally flat surface faces south, may not receive enough sunlight to justify investment in a solar unit. This chapter discusses orientation, vegetation, and ways to determine whether you have enough sunlight to make good use of the sun.

Chapter 3: Estimating Costs and Savings

Almost any solar energy system will save energy, but saving money depends on a number of factors, many of which can only be estimated, like the amount of sunlight on a given day, the price of conventional energy, and the interest rate. In this chapter we outline the elements of cost in a solar energy system, and provide a method, including work sheets, to help you estimate whether a solar unit is justified for your situation.

Part Two, "A Buyer's Guide" assumes that you have decided to invest in a solar energy system. It is designed to help you select a system and installer, and to contract for purchase and installation.

Chapter 4: Choosing the Right Dealer/Installer

Selecting, sizing, and installing a solar hot water system is not a simple task, and selection of a dealer or installer who can do a good job is essential to your being satisfied with the system. This chapter suggests ways to locate solar companies, lists the basic qualifications of a good company, and offers help in selecting the firm best for you.

Chapter 5: Choosing the Right System

Though selection of a dealer may be more important than the choice of system, you need some knowledge of the features of the various systems. This chapter will help you raise the right questions as you discuss the types of systems offered with the dealers or installers you are considering.

Chapter 6: Warranties and Contracts

Purchasing a solar energy system is a major financial decision for most families, and this chapter outlines issues which must be considered in preparing and signing a contract or purchase order and in assuring that the system will perform as promised.

Part Three, "Protecting Your Investment" assumes that you have now selected a dealer and a system, and have arrived at an agreement for purchase and installation. Ways to improve your chance of getting and keeping what you paid for are discussed in this final section.

Chapter 7: Getting a Good Installation

Although you will have selected a dealer/installer in whom you have confidence, you should have some understanding of what constitutes good workmanship. In this chapter we discuss, briefly, certain points to watch during the installation. For more detailed information on system installations, we suggest you obtain a copy of the Installation Guidelines report cited in the Introduction.

Chapter 8: Living with Your Solar Energy System

Once installed correctly, solar water heaters should operate as designed. But these systems do require periodic servicing and maintenance, and may need repairs. This chapter discusses these items, and concludes with suggestions of what to do if you end up with a serious complaint about your system.

The Appendices discuss system performance and durability, and provide sources of additional information on solar energy and its applications.

Part One

Is
Solar Water
Heating
Right for You?

How Solar Water Heaters Work

Mechanically, solar water heaters for the home are not much different from the more familiar gas and electric water heaters. What is unique is their heat source: a few collector panels that absorb solar radiation and convert it to usable heat. These collectors can go on the roof, against a wall, or even on a separate support frame in the yard, as long as they are exposed to direct sunlight most of the day.

The Basic Components

Collectors

The typical "flat-plate" collector panel used in water heating is an insulated weathertight box containing a dark solar absorber plate under one or more transparent covers. The dark absorber soaks up heat from sunlight that passes through the cover

and then gives the heat up to a heat-transfer fluid flowing past or through the absorber. This fluid (water, a non-freezing liquid, or air) delivers its heat directly or indirectly to water stored in an insulated tank.

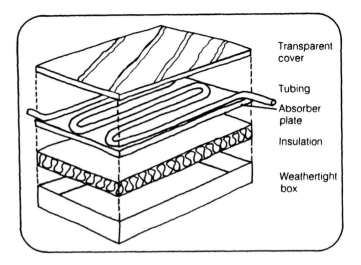

Transparent cover

Tubing

Absorber plate

Insulation

Weathertight box

Exploded view of collector

Storage tank

The solar-heated water may be stored:

in a tank that also houses an electric backup heating element (a "one-tank" system), or

in a separate tank that feeds into the tank of a conventional gas or electric water heater (a "two-tank" system), or into a "side-arm" water heating coil in a hydronic heating system.

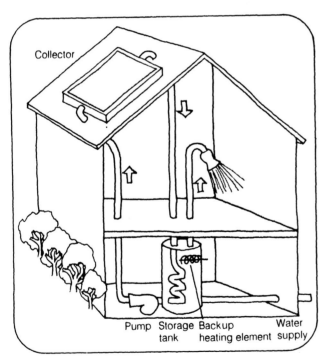

Collector

Pump Storage tank Backup heating element Water supply

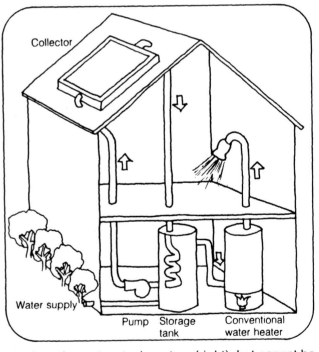

Collector

Water supply

Pump Storage tank Conventional water heater

A one-tank system (left) requires less space and usually costs less than a two-tank system (right), but cannot be used if the backup fuel is oil or gas. A two-tank system can deliver more hot water on a continuous basis.

Whether one or two tanks are used, solar energy really serves to *preheat* the household hot water. At night and on cloudy days, the conventional backup heater gives the water a boost to the desired temperature. However, on sunny days, when a typical solar unit is fully capable of raising water to 140°F (the maximum safe temperature setting for household water), the backup heater remains off. The solar storage tank is usually large enough to hold at least a day's supply of hot water.

Circulation and controls

Another element or component common to all but the simplest solar water heaters is some kind of circulation system to keep heat-transfer fluid continually moving from collectors to tank. Except where this flow is by natural convection, a pump is needed to force the fluid on its way. A control unit turns the pump on and off as temperature sensors dictate.

Some systems are equipped with devices that give an indication that the unit is operating. These gauges can be as simple as a signal light to let the owner know that the pump is running, together with dial thermometers on the collector inlet and outlet pipes and on the storage tank. If these are not supplied by the manufacturer, the installer can add them to the system at a relatively low cost.

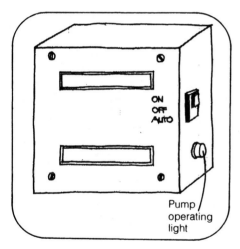

Pumps in solar units are typically very quiet. A signal light can be added to the control box to show when the pump is actually circulating fluid through the collectors.

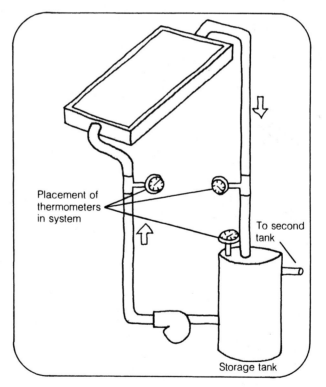

Two thermometers record the temperature of the fluid as it enters and as it leaves the collector. A third measures heat retention in the storage tank.

Common Types of Systems

All solar water heating systems can be characterized as either direct or indirect, depending on whether household water is heated directly in the collector or picks up the sun's heat indirectly.

In **direct** (open-loop) systems, the fluid heated in the collectors is plain water which flows directly to your faucet or washing machine.

In **indirect** (closed-loop) systems, the heat-transfer fluid is treated water, air, or some non-freezing liquid like an antifreeze solution or a special oil. The heat it picks up from the absorber plate is passed along to the house water through a heat exchanger, such as a coil inside the tank or wrapped around the storage tank shell under the insulation. (See illustration page 16).

Direct (Open-Loop) Systems

Thermosiphoning

The simplest direct system is the thermosiphoning water heater. Water circulates by natural convection and gravity, rising and falling in response to solar heat, just as air would. As long as the absorber keeps collecting heat, water warmed in the collector rises into a storage tank placed slightly above, while cooler tank water runs down to take its place.

Thermosiphoning units are simple, relatively inexpensive, and require little maintenance. In their simplest form they are not suited to freezing climates, because water remains in the collectors at all times. However, a valve can be added to drain the collectors when freezing temperatures occur.

Aside from the freezing potential, the greatest disadvantage to these systems is that the storage tank must be located at least two feet *above* the collectors. In many cases, this means finding a spot for a heavy water tank on an upper floor or under the roof ridge.

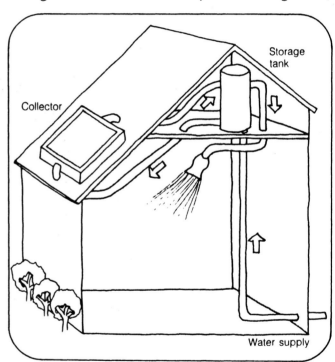

In a thermosiphoning system, cold water flows from bottom of tank to bottom of collector, and returns to the tank when warmed.

Pumped

A direct **pumped** system is often used when more flexibility in system layout is needed. With forced circulation, the tank need not be located above or even near the collectors. A direct pumped system typically relies on "drain-down" for protection against freezing. The pump moves water through the collectors only when there is enough solar radiation to produce useful heat. When the pump shuts off–whether by automatic control or due to power failure–the collectors are drained by gravity flow.

Direct systems, whether pumped or thermosiphoning, cannot be used in areas where water is hard or acidic. Scale deposits would quickly clog the inside of the absorber tubing, and corrosion can render a system inoperable.

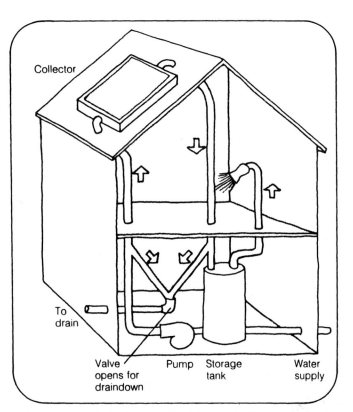

In a pumped draindown unit, solar-heated water flows to the storage tank for direct use by the household. When the pump shuts off, whatever water remains in the collectors drains away by gravity flow.

Indirect (Closed-Loop) Systems

The best choice for hard-water areas is an indirect, closed-loop system. The heat-transfer fluid never comes in direct contact with household water, so a corrosion-inhibiting solution can be circulated in the collector loop. Where there is a risk of freezing, an antifreeze solution or other non-freezing fluid is often used.

Some closed-loop systems prevent freezing by draining the collectors when the pump is not running. In this respect, these **drainback** systems function much like the open-loop "draindown" systems that have already been described above, except that the fluid is not dumped but drains back into an indoor holding tank.

In general, closed-loop systems permit the most flexible layouts and installations but are more expensive to purchase and install than open-loop systems.

15

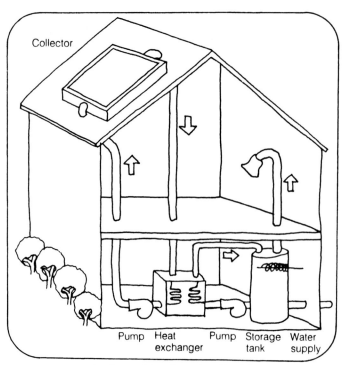

Indirect: Non-freezing liquid

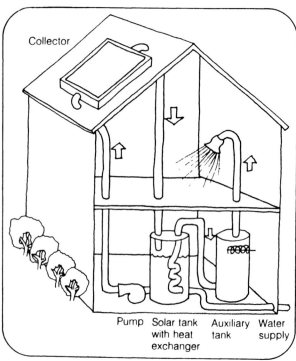

Indirect: Drain-back

At left, water from the storage tank is pumped through the heat exchanger, where it picks up heat from the solar-heated fluid flowing through. Because the collector fluid circulated in the system at right is not freeze-proof, the unit contains a "drainback" feature, meaning that the fluid drains into the solar tank when the pump shuts off.

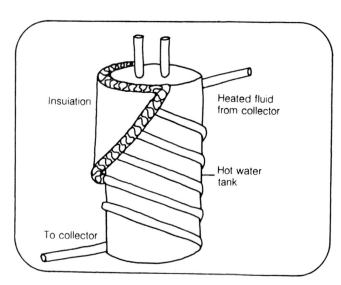

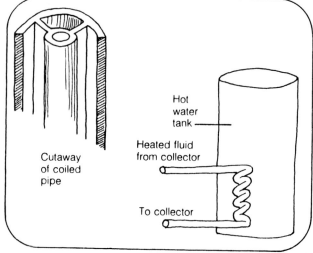

When a heat-transfer fluid contains additives that are toxic or non-drinkable, a "double-wall heat exchanger" greatly reduces the chance that this fluid could accidently mix with the house water supply. Both of the devices shown have two distinct walls or separations between fluid and water.

The next two chapters explore questions that many consumers will want to answer for themselves—at least in a preliminary way—before consulting a dealer or installer:

Is my house located where it can make good use of the sun?
(page 19)

Does solar water heating make economic sense for me?
(page 27)

Safety Issues

Solar hot water systems are not inherently "unsafe," but they consist of complex equipment. Certain safety issues must be considered in selecting, installing, and maintaining a solar system. Although they are not identified as "safety-related" items, the important points are discussed in this report:

Toxic transfer fluids: These usually require double-walled heat exchangers in a closed system; see pages 14 and 16.

Installation and maintenance access: The collector, storage tank, and controls should be located so that the installer and maintenance personnel can reach them safely; see page 24.

Protection from snow slides: In appropriate climates, snow and ice can build up on collectors, only to slide off when the collector warms. The collectors must be located or arranged so that sliding snow is not a safety hazard; see page 70.

Scalding: The maximum safe temperature for hot water systems is 140°F. Under good conditions, many collectors will heat the water to higher temperatures. A mixing valve can protect against scalding; see page 71.

Other safety issues are discussed in HUD's publication *Installation Guidelines for Solar DHW Systems.*

Making Good Use of the Sun

Chapter 2

After reading about how solar water heaters work, you may be wondering how you would fit such a system into your present house or make the best use of one in a planned new house. It makes sense to consult an experienced dealer or installer before making any detailed or firm plans. However, look around first to see if your home is located where it can make generally good use of the sun. Keep the following points in mind:

Although some buildings enjoy better exposure to the sun than others, most are adaptable to installation of a solar unit. Usually the amount of modification needed and the difficulty involved are relatively minor. However, in some cases modifications can add sizeably to costs.

In heavily developed areas or where there is dense shade from many tall overhanging trees, there may be no reasonable way to collect enough sunshine to run a solar energy system. This situation is usually quite obvious to the owner making a quick survey. If any doubts remain, consult a reputable dealer.

This chapter is intended to alert you to features of your house and site that could have an effect on your solar buying decision. The first section explains basic requirements for collecting solar radiation. The following section suggests guidelines to use in checking a house and/or site for:

exposure to sunlight throughout the day;

possible areas for mounting solar collectors; and

installation demands and restrictions.

Basic Requirements

A solar energy system needs to be placed so that plenty of sunshine falls onto its collection surface. The greatest amount strikes collectors that are:

aimed true south ("Orientation");

tilted up at right angles to the sun ("Tilt"); and

unshaded, especially when the sun's rays are most intense ("Shading," page 22).

Orientation

Orientation refers to the position of a surface relative to *true* south. Although collectors that face within 15° of true south receive the most sunshine, any unobstructed, generally south-facing surface is a potential collector location.

In many areas, a slightly westerly orientation is preferable to due south, both to avoid morning haze and to take advantage of the afternoon's higher outdoor temperatures, which lead to better collector performance.

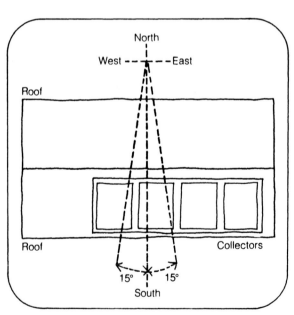

Tilt

Collector surfaces tilted at right angles to the sun's rays catch the most sunshine per unit area. An angle equal to local latitude is the closest ap-

20

proximation to that tilt or slope on a year-round basis. This means that the ideal roof for mounting solar collectors is pitched about the same number of degrees as local latitude. The *exact* tilt of a collection area is not crucial; a variation of 10 degrees one way or the other to suit a roof's pitch makes almost no difference.

An experienced installer can usually devise a means of compensating for a roof that is too low-pitched. On a flat or low-sloped roof, a support frame can be constructed to raise collectors to the recommended angle. However, there are several potential drawbacks to a free-standing collection area on a roof. If it protrudes far above the roof line, it can be an eyesore. It may interfere with chimney and plumbing vent clearances, requiring costly extensions to chimney or vents. And the frame must be sturdy enough to resist strong winds. An alternative is to specify one or more additional collector panels. Any such option will add to system cost.

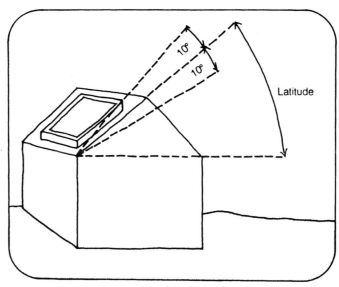

The ideal roof for mounting solar collectors is pitched within 10° either way of local latitude.

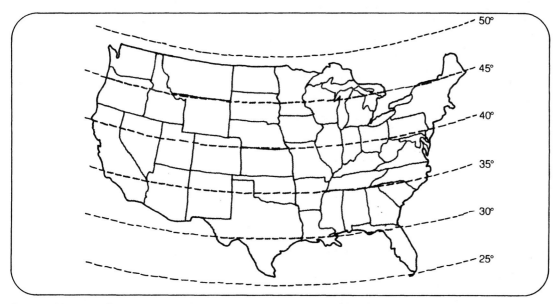

Check this map for the approximate latitude of your locale. Recommended collector tilt falls within 10° either way. For example, in Denver, Colorado, where the latitude is 40°, collectors tilted at any angle between 30° and 50° are within the recommended range.

21

Shading

A properly oriented and tilted collector that is mostly shaded between 9 a.m. and 3 p.m. will not produce much heat. Large, solid shadows greatly reduce output; small shadows and those cast early and late in the day make less difference in system performance. Types of shading to watch for are described in the next section ("Collector location," below).

Checking the House and Site

Now that you are generally familiar with what is needed to make good use of the sun, you will want to check your present home or future house site for any obstructions or restrictions that could prevent solar collectors from getting correct exposure to solar radiation. Things to consider include:

collector location;

access to sunlight;

installation complications; and

zoning laws and restrictive covenants.

Any of these would hamper—and some might rule out—the installation of a solar energy system on your property.

Collector location

The first thing to do is locate south. Though precision is not required at this point, you can easily find true south—which can vary by 20° from the magnetic south of the compass—by checking the shadow cast at solar noon by a stick held vertically. Solar noon occurs exactly midway between sunrise and sunset—times that your local newspaper probably publishes. This shadow will be a true north-south line.

Now look around for a generally unshaded, south-facing surface that is large enough for 40 to 100 square feet of collectors. (The right number of panels

When the roof on a house faces east and west, a freestanding rack can be used to tilt collectors against the south wall at the preferred angle. Collectors mounted vertically on the wall will not collect as much energy per square foot.

can be determined later.) If your home has no appropriate roof area, there may be good mounting surfaces on its south wall (see illustration) or on an adjacent shed, garage, or carport. Ground-mounted collectors on a support frame are another possibility. However, the use of a mounting frame or rack adds construction costs.

Access to sunlight

Shading is most likely to come from two sources:

Parts of the house itself. Chimneys, dormers, overhangs, and other elements can partially shade adjacent roof-mounted collectors.

Buildings or trees to the south. Obstructions like buildings or large evergreens that cause no interference in summer may cast long shadows when the winter sun is low in the sky. Trees that shed their leaves in the fall are not a problem if the summer sun can pass over them, since their leafless branches may not be much of a hindrance to the winter sun.

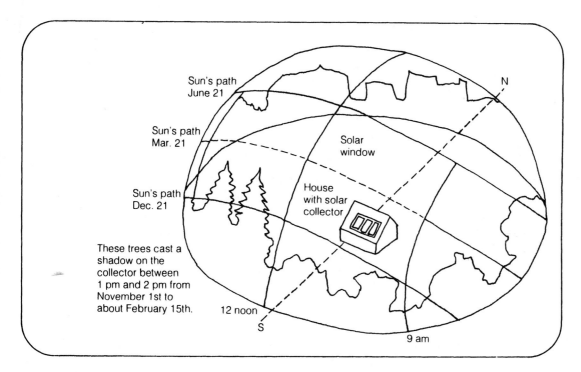

A limited amount of summer shading will be offset by the longer collection hours in summer and the higher temperatures then.

Though your house or site may be unshaded now, consider the possibility that growing trees or future construction on neighboring properties could block the sun in years to come. Land use regulations generally do not guarantee access to sunlight, though private agreements can sometimes be worked out among neighbors.

Installation complications

In a few cases obstructions to sunlight provide less of an obstacle to a solar installation than certain limitations of an existing house. Here are some questions to consider now and possibly discuss later with an installer, particularly with respect to costs:

Piping or ductwork. How difficult will it be to route pipes or ducts from the basement or ground floor to the roof? Will sections of wall or floor need to be torn out?

Storage tank. Is there room in your basement or ground floor for a solar storage tank that could be up to three feet in diameter and six feet tall? Can it be fitted in near the existing water heater? Can it be brought through existing stairways and doorways?

Working conditions. If you plan a roof-mounted collector, is there easy access to the attic? What about the slope and accessibility of your roof? Could an installer easily work there?

Roofing material. Can the collector supports be readily fastened to the roof? Slate and clay tile, which are brittle and chip very easily, are examples of materials requiring unusual care.

Aesthetics. How much will you be altering the appearance of your house? Will you be satisfied with the way it looks? Are your neighbors likely to object?

Zoning laws and restrictive covenants

If you decide to go ahead with a solar energy system, you will expect your installer to check into all applicable code and zoning regulations, just as he would do on any home-improvement project. In the meantime, make a preliminary check to find out if there are zoning restrictions in your community that could prevent you from placing solar equipment outside of your home or limit where it can be installed.

Ordinances specifying height limits, front- and side-yard setbacks, and lot density could apply directly to your planned system. Ground-mounted collectors might be expected to meet all requirements of a separate building, or could violate setback requirements. Even roof-mounted collectors can stick out too far.

In some communities, building plans are subject to the aesthetic judgment of an architectural review board. Restrictions on building appearance may also apply if you live in an historic district or if you signed a restrictive covenant when you purchased your home.

None of these is necessarily a permanent barrier. Variances are frequently granted for good cause, and the use of energy-conserving equipment may well qualify.

Whether solar water heating makes sense for you depends not only on the location of your house and/or site, but also on how the economics work out in your situation. Ways of estimating costs and savings are explained in the next chapter.

Chapter 3

Estimating Costs and Savings

Solar water heating systems generally cost $2000 to $3000, installed (as of 1979). The cost can be reduced in several ways, most notably by the Federal tax credit, which can amount to as much as 40 percent of the total. (This credit and other ways to cut costs are discussed later in the chapter, on page 31.)

Once the system is installed, its cost is offset over time by monthly savings on fuel bills. A correctly designed and installed system can save 50 percent or more of your annual bill for hot water.

Differing expectations

Of course, not all motives for buying a solar water heater are directly related to costs and savings. Often solar buyers want greater self-reliance.

Some are intrigued with making use of the sun's energy; others hope to help conserve fossil fuels, cut down on pollution, or gain insurance against shortages. However, for most consumers the desire to save on future fuel bills is a central, if not the primary, motivation.

Ways of looking at solar economics are as numerous and varied as the reasons for using solar energy. Let's say your main reason is something other than economic. In that case, you could be less concerned with net costs and savings than with simply obtaining financing. Or you could be focusing on low first costs instead, content to let savings accrue as they will. Even supposing that your motivation is strictly economic—that is, you plan to buy a solar energy system only if it will improve your economic situation—there is no guarantee that you will agree with someone else on what constitutes an improvement. For example, consider the different perspectives of three consumers asking, "Will hot water from the sun save me enough to make it worthwhile?":

The first expects a solar unit to pay off its initial cost with fuel bill savings within a few years.

Another, impressed with the marketability and resale value added to his house, will wait happily for a longer period.

Still a third is looking for an annual profit, expecting fuel bill savings to pay a better return on the cash invested than a savings account would. *If* the solar unit is covered by a loan, he may expect fuel bill savings to be greater than the loan payments, at least after the first few years.

Because expectations vary so widely, a final decision on what is economic must be left to the individual. The next two sections identify costs associated with solar water heating (below) and discuss ways to reduce them (page 31). The final section (page 33) provides some useful tools for assessing the economic impacts of solar energy use, and illustrates a simple two-step method of comparing estimated first costs and fuel bill savings. The examples relate the results to economic benefits that go way beyond fuel savings.

Identifying Costs

The total cost of a solar water heater will of course exceed the price of the packaged kit or assembly. Materials and labor needed for installation may add at least half again to the cost of the equipment. In new construction, the collectors and pipes or ducts can be built into the roof and walls, making installations in new homes less expensive than others. If there is difficulty making way for the pipe runs or the added storage tank in an existing home, installation costs will be higher than usual.

Sizing

Solar water heating units are generally planned, or "sized," to be large enough to provide *one day's* hot water needs under *average* conditions. This does not mean that the typical system can supply all the hot water drawn on any day regardless of the weather or other circumstances. Enough collectors to heat all the water a family might use during the darkest days of winter would cost far more than could ever be saved on fuel bills, and a large array of collectors would be underused much of the time.

Over the entire year, a well-sized solar unit can be expected to meet 50 to 75 percent of a household's demand for hot water (with the backup system providing the rest). When conditions are favorable, as during the sunny summer months, the solar unit alone can heat almost all the hot water used.

The actual solar collector area needed for a given installation depends on a number of variables and you may wish to rely on the recommendation of a qualified dealer or contractor. However, for purposes of initially estimating costs and savings, you can reasonably consult TABLE C, below, to find a collector size for a household like yours.

TABLE C. Suggested Collector Sizes For Solar Water Heating Systems

Note: The collector area indicated in each case can be expected to provide, at the lowest possible cost per unit of heat produced, at least 50 percent of the hot water used annually by the average household of that size. A collector panel typically has 20 square feet of collecting surface.

Location	Number of Users 2	4	6	Location	Number of Users 2	4	6
	Square Feet				Square Feet		
BIRMINGHAM, AL	40	60	80	BILLINGS, MT	40	60	80
FAIRBANKS, AK	60	80	100	GREAT FALLS, MT	40	60	80
TUCSON, AZ	40	60	60	LINCOLN, NE	40	60	80
LITTLE ROCK, AR	40	60	80	LAS VEGAS, NV	40	60	60
LOS ANGELES, CA	40	60	80	RENO, NV	40	60	80
SACRAMENTO, CA	40	60	60	CONCORD, NH	60	80	100
SAN FRANCISCO, CA	40	60	80	ATLANTIC CITY, NJ	40	60	80
DENVER, CO	40	60	80	ALBUQUERQUE, NM	40	60	60
GRAND JUNCTION, CO	40	60	80	ALBANY, NY	60	80	100
HARTFORD, CT	60	80	100	NEW YORK, NY	60	80	100
WILMINGTON, DE	60	80	100	ROCHESTER, NY	60	80	100
WASHINGTON, DC	60	80	100	SYRACUSE, NY	60	80	100
JACKSONVILLE, FL	40	60	80	CAPE HATTERAS, NC	40	60	80
MIAMI, FL	40	60	80	RALEIGH, NC	40	60	80
TALLAHASSEE, FL	40	60	80	BISMARCK, ND	40	80	100
TAMPA, FL	40	60	80	CLEVELAND, OH	60	80	100
ATLANTA, GA	40	60	80	COLUMBUS, OH	60	80	100
SAVANNAH, GA	40	60	80	OKLAHOMA CITY, OK	40	60	80
HILO, HI	40	60	80	TULSA, OK	40	60	80
HONOLULU, HI	40	60	60	MEDFORD, OR	40	60	80
BOISE, ID	40	60	80	PORTLAND, OR	60	80	100
POCATELLO, ID	40	60	80	PHILADELPHIA, PA	60	80	100
CHICAGO, IL	60	80	100	PITTSBURGH, PA	40	80	80
PEORIA, IL	40	60	80	STATE COLLEGE, PA	60	80	100
INDIANAPOLIS, IN	60	80	100	PROVIDENCE, RI	60	80	100
DES MOINES, IA	40	80	80	CHARLESTON, SC	40	60	80
WICHITA, KS	40	60	80	RAPID CITY, SD	40	60	80
LEXINGTON, KY	60	80	100	NASHVILLE, TN	40	60	80
LOUISVILLE, KY	60	80	100	AMARILLO, TX	40	60	80
NEW ORLEANS, LA	40	60	80	DALLAS, TX	40	60	80
SHREVEPORT, LA	40	60	80	EL PASO, TX	40	60	60
CARIBOU, ME	60	80	100	HOUSTON, TX	40	60	80
PORTLAND, ME	60	80	100	SALT LAKE CITY, UT	40	60	80
BALTIMORE, MD	60	80	100	BURLINGTON, VT	60	80	100
AMHERST, MA	60	80	100	MT. WEATHER, VA	60	80	80
BOSTON, MA	60	80	100	NORFOLK, VA	40	60	80
LANSING, MI	60	80	100	RICHMOND, VA	60	80	80
SAULT STE. MARIE, MI	60	80	100	SEATTLE, WA	60	80	100
MINN.-ST. PAUL, MN	60	80	100	SPOKANE, WA	40	60	80
JACKSON, MS	40	60	80	CHARLESTON, WV	60	80	100
KANSAS CITY, MO	40	60	80	MADISON, WI	60	80	100
ST LOUIS, MO	40	60	80	CASPER, WY	40	60	80
				CHEYENNE, WY	40	60	80

This table is repeated in larger type on page 117.

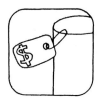

Getting an estimate

To compare probable costs and savings by the method explained and illustrated later in this chapter (page 34), you will need an estimate of system cost. Since you may wish to postpone getting actual estimates until a later date when you feel better prepared to discuss prices intelligently with a dealer/installer, you can use TABLE D (below) to find what solar purchasers have been paying for an installed system using a collector area appropriate for a household of your size. The table shows installed costs that were typical in 1979 for both new and existing homes.

When you do approach a dealer for an estimate, ask for the price of a system including all components that will provide between 50 and 75 percent of the hot water used over a year by the average household of your size in your location. The installer's estimate should specify everything he will supply and do, and should be based on an inspection of your home.

TABLE D. Typical Costs In 1979 For Solar Hot Water Systems

Note: This table, which is based on information from the HUD solar demonstration program and from industry sources, lists approximate installed costs, *before* tax credits, that were typical in 1979 in areas where freeze protection is necessary. Somewhat lower costs may be expected in the warm southern states. If you need help in estimating the collector size best suited to your hot water use, refer to Table C. Areas given below are multiples of 20 square feet, the average size of a collector panel.

Collector Size (Sq. Ft.)	Installed Cost For New Homes $	Installed Cost For Existing Homes $
20	1450	1600
40	1900	2050
60	2300	2550
80	2750	3000
100	3200	3500

This table is repeated in larger type on page 118.

Ways to Reduce Costs

The most effective ways to reduce costs are to:

adopt some simple conservation measures;

take advantage of income tax credits and other tax incentives;

install your own system.

Conservation

There are several ways to reduce water heating demand without inconvenience or discomfort.

Install flow restrictors in shower heads and faucets, which effectively reduce the amount of hot water drawn.

Add more insulation around the existing water tank. Ready-made insulating jackets are available in many hardware and building supply stores.

Experiment with reduced water-heater settings. If the thermostat is now at 140°, you can save seven percent on your utility bill by reducing it to 120°, which is hot enough for most uses.

If you have an electric water heater, investigate timers limiting hours of service. These are cost-effective where electricity consumed during off-peak hours is billed at a lower rate. Timers (similar to set-back space heating thermostats) can be set to provide hot water only when it is needed and can save electricity.

Tax breaks

To encourage energy savings, the Federal government as well as some state and local governments now offer cost-reducing incentives in the form of tax credits, deductions, and exemptions:

On the Federal level, the Internal Revenue Service allows an energy tax credit that can be subtracted from income tax owed.

The credit is 40 percent of the first $10,000. (Because the credit applies to solar *space* heating systems as well, the credit ceilings go well beyond the costs associated with a solar *water* heating system only.) To find out how much this would reduce your estimated system cost, see TABLE E (page 32).

Various states offer similar credits and deductions from state income taxes. Many permit solar units to be assessed for property tax pur-

TABLE E. Federal Energy Tax Credits

Note: To find the credit on values not appearing in the table, follow this simple method:

If the installed cost is $10,000 or less, multiply it by 40 percent.

(For a copy of I.R.S. Publication 903, "Energy Credits for Individuals," return the request form at the back of the guide.)

If Your Installed Solar Hot Water System Cost Is: $	Then Your Tax Credit Is: $
1000	400
1250	500
1450	580
1600	640
1750	700
1900	760
2050	820
2300	920
2550	1020
2750	1100
3000	1200
3200	1280
3500	1400

This table is repeated in larger type on page 118.

poses at less than original value. Some exempt solar equipment from sales tax. To find out which tax incentives are available to you, write or call the National Solar Heating and Cooling Information Center.

Do-it-yourself

Any work that you can do on your own water heating system will of course cut costs. But, in the long run, installing it yourself will save you money *only* if you have the expertise to do it well. A few seemingly minor mistakes could cause costly damage or even ruin the system.

If you believe that you have the time and superior skills necessary for an installation, decide which kit you might purchase and write for a copy of the assembly instructions. Then re-evaluate your skills and your commitment. Remember that:

You assume all risks involved.

Warranties usually depend on installation being done in a "workman-like manner," according to manufacturer's recommendations.

The final hook-ups may have to be made by a licensed plumber and electrician.

Comparing Costs and Savings

The clear economic benefit of a solar energy system is obviously the money saved each month on fuel bills. At first these savings may not be large, but they will grow along with fuel prices, which are rising very rapidly. (A worksheet for estimating your possible fuel bill savings appears on page 105.)

Payback

Consumers sometimes ask about "payback," that is, they want to know when savings from a solar unit will pay back its cost, plus any interest. Some shoppers will not consider a solar device that would not pay for itself in four or five years (though they might not place the same demand on other useful devices or amenities in the home). Since payback estimates ignore the appreciable intrinsic value of a solar energy system, many experts believe they should not be the only basis for comparison.

Improved cash flow

If you expect solar energy to increase the amount of money you have left after paying your monthly bills, you are looking, in economic terms, for improved "cash flow." With *cash* purchases, several years may pass before total fuel bill savings equal your initial outlay, but your monthly utility bill will be cut from the outset, so improved cash flow begins at once. You can expect to recoup your cash investment when the house is sold.

When *financing* is involved, favorable cash flow will occur as soon as monthly fuel bill savings come to more than monthly loan costs. When the solar loan can be paid back in small increments over a long period of time, as in the case of mortgage loans—typically 25 to 30 years—cash flow may improve early. It is not unusual for a solar unit to reduce fuel bills to the point where the owner's energy savings are greater than the solar portion of the mortgage payment.

Because monthly payments are substantially higher on short-term loans, such as those available for home improvements, and on extended-payment credit-card purchases, improved cash flow will probably occur only after the loan is repaid. From that time on, however, the money saved on monthly fuel bills will be available for other purposes. Savings keep growing as long as the system lasts—15 or 20 years, if it is well made.

A better "savings account"

Cash invested in a solar energy system often earns a better return (in *tax-free* utility bill savings) than it would in a savings account. Consider, for example, that $2,000 on deposit in a conventional savings account (paying, say, 5¾ percent) will probably return no more than 4 percent, or $80 a year, after income taxes. A solar water heater costing the same amount and supplementing an electric unit could cut the average family's electric bill at least $10 a month, or $120 a year—netting 50 percent more than the savings account.

However, keep in mind that the solar investment can be converted to cash only when the house is sold.

Added value and marketability

Not enough solar-equipped houses have changed hands to produce the kind of market data that helps real estate experts pin down the value that a solar unit adds to property. But common sense suggests that any system that cuts utility bills year after year is bound to enhance the value of a house as the fuel situation gets worse.

Economic worksheets

The worksheets at the back of this guide (page 103) have been prepared to help you understand various economic impacts and their implications for you. By looking up a few numbers in tables and doing some simple arithmetic, you can estimate fuel bill savings and compare them with approximate costs. Exact predictions are not possible, since actual savings depend on fuel prices, which are escalating at an unpredictable rate. In addition, one household's habits of using hot water may differ markedly from average use patterns. However, the figure you come up with should shed some helpful light on your own particular solar economics—whether you:

are thinking of a solar water heater for a **new home** being planned, or

might add one to an **existing house,** and

whether you intend to:

include the cost in your **mortgage loan,**

pay **cash,** or

take out a **short-term loan.**

Each of the four blank worksheets is organized as a separate step. A prospective solar buyer will need to complete only *two* of the four. Everyone should complete STEP 1 (page 105), which provides a quick estimate of the annual fuel bill savings that a particular household could expect from a solar water heater. The remaining three steps are designed to compare these savings to costs. Your choice of which to complete depends on how you would pay for a solar unit:

For mortgage loans, complete STEP 2 (page 106). This step compares costs and savings on the basis of improved monthly cash flow, that is, how long it will take for monthly energy savings to exceed loan costs.

For cash purchases, complete STEP 3 (page 107). This step compares costs and savings on the basis of how long it will take for solar savings to recover the initial cost of the system less any tax credits.

For short-term loans (personal, home-improvement, or energy-conservation) and credit-card purchases, complete STEP 4 (page 108). This step compares costs and savings on the basis of how long it will take for solar savings to recover the financed cost of the system, that is, the initial cost less any tax credits plus all interest.

To illustrate the simple completion of the worksheets, three examples have been fully worked on the following pages.

Example 1 (page 36) illustrates projects financed with mortgage loans (STEP 2) and using electrical backup systems.

Example 2 (page 40) illustrates projects purchased with cash (STEP 3) and using oil as a backup energy source.

Example 3 (page 44) illustrates projects financed with short term loans (STEP 4) and using gas as a backup energy source.

Appendix E outlines the technical assumptions which have been used to develop the worksheets and tables.

EXAMPLE 1

A family in the **Midwest** who

● will include the cost of a solar water heater in the **mortgage** on their new home

● will use **electricity** for backup water heating

In their new home in Kansas City, Missouri, the Smiths, a family of **four,** plan to include a solar water heater with **60 square feet** of collector area (from TABLE C). The cost of the installed unit will be covered in their **30-year, 11½%** mortgage. An electric element will provide backup heating.

STEP 1.   Estimate Fuel Bill Savings

Procedure

1a. In TABLE A (pages 110-115). find the number of units of electricity (page 110). natural gas (page 112). or oil (page 114) that your solar water heater would save annually. Fill in here.

3300 KWH	**ANNUAL ENERGY SAVINGS**

1b. Fill in the current Unit Price of that fuel, and MULTIPLY.

$\times$

$.06/KWH	**UNIT PRICE OF FUEL**

You can expect a solar water heater to cut the first year's fuel bills about this much.

$=$

$ **198**	**FIRST-YEAR FUEL BILL SAVINGS**

OPTIONAL

1c. To estimate fuel savings over a longer period, refer to TABLE B (page 116). Find the cumulative savings factor for the period of years that you are considering. Fill in here, and MULTIPLY.

$\times$

27.2	**CUMULATIVE SAVINGS FACTOR**

Over that time period, a solar water heater can cut your fuel bills about this much.

$=$

$ **5,386**	**LONG-TERM FUEL BILL SAVINGS**

Blank form for this step on page 105

To find out when their solar unit will start saving more on fuel bills than it adds to their monthly mortgage payments, the Smiths need to do one more step. **Following worksheet instructions, they go on to STEP 2.**

TABLE A. Annual Energy Savings From A Typical Solar Water Heater

1. Kilowatt-Hours Of Electricity Saved

Number of Occupants	2		4		
Collector Area (sq. ft.)	40	60	40	60	80
Location					
MINNESOTA					
MINN-ST. PAUL	2000	2500	2300	3100	3800
MISSISSIPPI					
JACKSON	1900	2300	2400	3100	3600
MISSOURI					
KANSAS CITY	2100	2500	2500	3300	4000
ST. LOUIS	2000	2400	2400	3200	3800
MONTANA					
BILLINGS	2300	2700	2700	3700	4300
GREAT FALLS	2200	2700	2600	3500	4200

Full table on page 110

From TABLE A: The Smiths' probable annual energy savings in Kansas City would be 3300 kilowatt-hours. At their 1979 year-end electric rate of 6¢/kWh.

3300 × .06 = $198.00

estimated first-year fuel bill savings would be $198.

TABLE B. Cumulative Savings Factors

Time Period (years)	Fuel Price Escalation Rate (% per year)					
	4	6	8	10	12	14
1	1.0	1.0	1.0	1.0	1.0	1.0
5	5.4	5.6	5.9	6.1	6.4	6.6
10	12.0	13.2	14.5	15.9	17.5	19.3
15	20.0	23.3	27.2	31.8	37.3	43.8
20	29.8	36.8	45.8	57.3	72.1	91.0

Full table on page 116

From TABLE B: The Smiths expect to live in their new home at least 15 years, and think the price of backup electricity will escalate at an average annual rate of 8%. Multiplying, therefore, by a cumulative savings factor of 27.2,

$198 × 27.2 = $5,385.60

they find their system could save almost $5,400 during that period.

EXAMPLE 1 (continued)

After estimating, in STEP 1, how much a solar water heater will cut their fuel bills, the Smiths need do only one more step to find out how soon those monthly savings could be expected to exceed the system's loan costs.

STEP 2. Compare Costs and Savings— Mortgage Loans

Procedure

2a. From TABLE D (page 118), or from your dealer's estimate, fill in your Installed System Cost.

$ 2,300	**INSTALLED SYSTEM COST**

2b. From TABLE E (page 118), fill in the Federal Energy Tax Credit on this cost, and SUBTRACT.

− $ 920	**ENERGY TAX CREDIT**

This is what your system will cost you after you claim the Federal tax credit. (See note below on state tax credits.)

= $ 1,380	**NET SYSTEM COST**

2c. From TABLE F (page 119), select the annual loan payment factor corresponding to your mortgage's probable interest rate and term. Fill in here, and MULTIPLY.

× .119	**LOAN PAYMENT FACTOR**

This is the portion of your annual mortgage payment that covers your solar water heating system.

= $ 164	**ANNUAL SOLAR LOAN COST**

2d. From STEP 1, fill in First-Year Fuel Bill Savings, and DIVIDE.

÷ $ 198	**FIRST-YEAR FUEL BILL SAVINGS**

This is the ratio of your annual solar loan cost to first-year savings.

= .83	**LOAN COST/ SAVINGS RATIO**

2e. From TABLE G (page 119), fill in the year during which annual fuel savings will exceed annual loan costs and create a favorable cash flow.

IMMEDI- ATELY	**FAVORABLE CASH FLOW BEGINS**

Blank form for this step on page 106

The Smiths decide to proceed with a solar water heating system. By having the unit installed as their new house is built, rather than added on later, they are covering it under the mortgage and keeping costs down. Their annual solar loan cost of $180 is well under estimated first-year fuel bill savings.

TABLE D. Typical Costs In 1979 For Solar Hot Water Systems		
Collector Size (Sq. Ft.)	Installed Cost For New Homes $	Installed Cost For Existing Homes $
20	1450	1600
40	1900	2050
60	2300	2550
80	2750	3000

Full table on page 118

From TABLE D: The Smiths' installed system cost is an estimated $2300.

TABLE E. Federal Energy Tax Credits	
If Your Installed Solar Hot Water System Cost Is: $	Then Your Tax Credit Is: $
2050	820
2300	920
2550	1020

Full table on page 118

From TABLE E: Deduction of the allowable credit of $920

$2300 − $920 = $1380

yields a net system cost of $1380.

TABLE F. Annual Loan Payment Factors For Mortgage Loans								
Term of Loan (Years)	Interest Rate (%)							
	9	9½	10	10½	11	11½	12	13
20	.108	.112	.116	.120	.124	.128	.132	.141
25	.101	.105	.109	.113	.118	.122	.126	.135
30	.097	.101	.105	.110	.114	.119	.123	.133

Full table on page 119

From TABLE F: Multiplying by the loan payment factor corresponding to their 30-year 11½% mortgage,

$1380 × .119 = $164.22

the Smiths find their solar unit will add about $164 to their annual mortgage payments. When this solar loan cost is divided by first-year fuel bill savings (from STEP 1),

$164 ÷ $198 = .83

a loan cost-to-savings ratio results.

TABLE G. Year In Which Favorable Cash Flow Begins						
Loan Cost/ Savings Ratio	Fuel Price Escalation Rate					
	4	6	8	10	12	14
1.0*	1	1	1	1	1	1
1.1	4	3	3	2	2	2

*If the ratio is less than 1.0, favorable cash flow begins immediately.

Full table on page 119

From TABLE G: Since the ratio is less than 1.0, the Smiths discover that their solar unit would immediately cause favorable cash flow.

EXAMPLE 2

A family in the **Northeast** who

- will pay **cash** for a solar water heater

- now heats water with **fuel oil**

The Joneses of Hartford, Connecticut, are thinking of adding solar water heating to cut down on the use of their oil-fueled heater. For their family of **four, 80 square feet** of collector area is suggested (TABLE C, page 117);

STEP 1. Estimate Fuel Bill Savings

Procedure

1a. In TABLE A (pages 110-115), find the number of units of electricity (page 110), natural gas (page 112), or oil (page 114) that your solar water heater would save annually. Fill in here.

150 gal.	ANNUAL ENERGY SAVINGS

1b. Fill in the current Unit Price of that fuel, and MULTIPLY.

× $ **.92**	UNIT PRICE OF FUEL

You can expect a solar water heater to cut the first year's fuel bills about this much.

= $ **138**	FIRST-YEAR FUEL BILL SAVINGS

OPTIONAL

1c. To estimate fuel savings over a longer period, refer to TABLE B (page 116). Find the cumulative savings factor for the period of years that you are considering. Fill in here, and MULTIPLY.

× **17.5**	CUMULATIVE SAVINGS FACTOR

Over that time period, a solar water heater can cut your fuel bills about this much.

= $ **2,415**	LONG-TERM FUEL BILL SAVINGS

Blank form for this step on page 105

Following worksheet instructions, the Joneses go on to STEP 3—to find out how long it would take for fuel bill savings to equal the system's initial cost.

TABLE A. Annual Energy Savings From A Typical Solar Water Heater

3. Gallons Of Fuel Oil Saved

Number of Occupants	2		4		
Collector Area (sq. ft.)	40	60	40	60	80
Location					
CALIFORNIA					
LOS ANGELES	110	130	140	180	210
SACRAMENTO	110	120	140	180	200
SAN FRANCISCO	110	130	140	180	210
COLORADO					
DENVER	140	150	180	230	250
GRAND JUNCTION	130	140	160	210	240
CONNECTICUT					
HARTFORD	80	100	90	130	150
DELAWARE					
WILMINGTON	90	110	110	140	170

Full table on page 114

From TABLE A: The Joneses' probable annual energy savings in Hartford is 150 gallons of fuel oil. At the 1979 year-end price of 92¢/gallon,

150 × .92 = $138.00

estimated first-year fuel bill savings would be about $138.

TABLE B. Cumulative Savings Factors

Time Period (years)	Fuel Price Escalation Rate (% per year)					
	4	6	8	10	12	14
1	1.0	1.0	1.0	1.0	1.0	1.0
5	5.4	5.6	5.9	6.1	6.4	6.6
10	12.0	13.2	14.5	15.9	17.5	19.3
15	20.0	23.3	27.2	31.8	37.3	43.8
20	29.8	36.8	45.8	57.3	72.1	91.0

Full table on page 116

From TABLE B: The Joneses expect to stay on in their present house about 10 years, and foresee a 12% increase in fuel prices each year. Multiplying, therefore, by a cumulative savings factor of 17.5

$138 × 17.5 = $2415.00

they find that a solar water heater could cut fuel bills about $2415 during their remaining years in the house.

EXAMPLE 2 (continued)

After estimating, in STEP 1, how much a solar water heater will cut their fuel bills, the Joneses need do only one more step to find out how many years it would take for the solar unit to return its initial cost.

STEP 3. Compare Costs and Savings— Cash Purchases

Procedure

3a. From TABLE D (page 118), or from your dealer's estimate, fill in your Installed System Cost.

$ **3,000**	**INSTALLED SYSTEM COST**

3b. From TABLE E (page 118), fill in the Federal Energy Tax Credit on this cost, and SUBTRACT.

− $ **1200**	**ENERGY TAX CREDIT**

This is what your system will cost you after you claim the Federal Tax Credit. (See note below on state tax credits.)

= $ **1800**	**NET SYSTEM COST**

3c. From STEP 1, fill in First-Year Fuel Bill Savings, and DIVIDE.

÷ $ **138**	**FIRST-YEAR FUEL BILL SAVINGS**

This is the ratio of your system's net cost to first-year savings.

= **13**	**COST SAVINGS RATIO**

3d. From Table H (page 120), fill in the year during which total fuel savings will exceed initial costs.

IN THE **9th** YEAR	**COSTS RETURNED**

Blank form for this step on page 107

Fuel bill savings will probably return the cost of a solar water heating system during the 10 years the Joneses expect to remain in their present home. In any case, since they believe that most, if not all, of that cost will be recovered when the house is sold, they look upon fuel bill savings as equivalent to tax-free interest on a savings account.

If the installed system cost is recovered in the selling price, the $3615 savings over 10 years ($1200 tax credit, plus estimated fuel bill savings of $2415) would be equivalent to a tax-free annual return of 8.2 percent.

42

TABLE D. Typical Costs In 1979 For Solar Hot Water Systems		
Collector Size (Sq. Ft.)	Installed Cost For New Homes $	Installed Cost For Existing Homes $
20	1450	1600
40	1900	2050
60	2300	2550
80	2750	3000
100	3200	3500

Full table on page 118

From TABLE D: The Joneses' installed system cost would be $3000.

TABLE E. Federal Energy Tax Credits	
If Your Installed Solar Hot Water System Cost Is: $	Then Your Tax Credit Is: $
2750	1100
3000	1200
3200	1280
3500	1400

Full table on page 118

From TABLE E: Deduction of the allowable credit of $1200

$$\$3000 - \$1200 = \$1800$$

yields a net system cost of $1800. When this cost is divided by first-year fuel bill savings (from Step 1) of $138,

$$\$1800 \div \$138 = 13.0$$

a cost-to-savings ratio of 13 results.

TABLE H. Year In Which Accumulated Savings Exceed Total Costs							
Cost/Savings Ratio	Fuel Escalation Rate (% per year)						
	4	6	8	10	12	14	16
10.	9	9	8	8	7	7	7
15.	12	12	11	10	10	9	9
20.	15	14	13	12	11	11	10
25.	18	16	15	14	13	12	11

Full table on page 120

From TABLE H: Using this ratio, the Joneses discover that if fuel prices rise as they expect—about 12% yearly—savings will return system cost in the 9th year.

EXAMPLE 3

A family in the **Southeast** who

- will pay for a solar water heater with a **home-improvement loan at 13%**

- now heats water with **natural gas**

The Browns have just moved into an older house in Raleigh, North Carolina, which they hope to renovate gradually over the next 10 to 15 years. Mindful of rising fuel prices, they plan to borrow the money for a solar water heating system to supplement their gas-fueled water heater. For their family of **four,** they will need **60 square feet** of collector (from TABLE C).

STEP 1. Estimate Fuel Bill Savings

Procedure

1a. In TABLE A (pages 110-115), find the number of units of electricity (page 110), natural gas (page 112), or oil (page 114) that your solar water heater would save annually. Fill in here.

170 therms	ANNUAL ENERGY SAVINGS

1b. Fill in the current Unit Price of that fuel, and MULTIPLY.

× | $ *.36* | UNIT PRICE OF FUEL

You can expect a solar water heater to cut the first year's fuel bills about this much.

= | $ *61* | FIRST-YEAR FUEL BILL SAVINGS

OPTIONAL

1c. To estimate fuel savings over a longer period, refer to TABLE B (page 116). Find the cumulative savings factor for the period of years that you are considering. Fill in here, and MULTIPLY.

× | *31.8* | CUMULATIVE SAVINGS FACTOR

Over that time period, a solar water heater can cut your fuel bills about this much.

= | $ *1940* | LONG-TERM FUEL BILL SAVINGS

Blank form for this step on page 105

Following worksheet instructions, the Browns go on to STEP 4—to find out how long it would take for fuel bill savings to equal the system's financed cost.

TABLE A. Annual Energy Savings From A Typical Solar Water Heater

2. Therms Of Natural Gas Saved

Number of Occupants	2		4		
Collector Area (sq. ft.)	40	60	40	60	80
Location					
NEW YORK					
ALBANY	110	140	130	170	210
NEW YORK	90	120	110	150	180
ROCHESTER	90	110	100	140	170
SYRACUSE	90	110	100	140	170
NORTH CAROLINA					
CAPE HATTERAS	110	140	140	180	210
RALEIGH	110	130	130	170	210
NORTH DAKOTA					
BISMARCK	130	160	150	200	250

Full table on page 112

From TABLE A: The Browns' probable annual energy savings in Raleigh would be 170 therms of natural gas. At the 1979 year-end price of 36¢/therm,

170 × .36 = 61.20

estimated first year fuel bill savings would be about $61.

TABLE B. Cumulative Savings Factors

Time Period (years)	Fuel Price Escalation Rate (% per year)					
	4	**6**	**8**	**10**	**12**	**14**
1	1.0	1.0	1.0	1.0	1.0	1.0
5	5.4	5.6	5.9	6.1	6.4	6.6
10	12.0	13.2	14.5	15.9	17.5	19.3
15	20.0	23.3	27.2	31.8	37.3	43.8
20	29.8	36.8	45.8	57.3	72.1	91.0

Full table on page 116

From TABLE B: The Browns believe they can count on natural gas prices to rise at least 10% annually, over the next 15 years. Multiplying, therefore, by a cumulative savings factor of 31.8,

$61 × 31.8 = $1939.80

they find that a solar water heater could save them about $1940 during the years they are raising their family.

45

EXAMPLE 3 (continued)

Among the dealers listed in the solar directory distributed by the North Carolina State Energy Office, the Browns have found one in Raleigh who will charge them $2050 to install a reputable packaged system.

STEP 4. Compare Costs and Savings— Short-Term Loans

Procedure

4a. From TABLE D (page 118), or from your dealer's estimate, fill in your Installed System Cost.

$ **2,050** — INSTALLED SYSTEM COST

4b. From TABLE E (page 118), fill in the Federal Energy Tax Credit on this cost, and SUBTRACT.

— $ **820** **513** **1333** — ENERGY TAX CREDIT

This is what your system will cost you after you claim the Federal Tax Credit. (See note below on state tax credits.)

= $ **717** — NET SYSTEM COST

4c. From TABLE J (page 121), select the loan payment factor corresponding to your loan term and interest rate, and MULTIPLY.

× **1.37** — SHORT-TERM LOAN FACTOR

This is the total amount you will pay out over the life of the loan, including interest.

= $ **982** — OVERALL FINANCED COST

4d. From STEP 1, fill in First-Year Fuel Bill Savings, and DIVIDE.

÷ $ **61** — FIRST-YEAR FUEL BILL SAVINGS

This is the ratio of your overall financed cost to first-year savings.

= **16** — LOAN COST/ SAVINGS RATIO

4e. From TABLE H (page 120), fill in the year during which total fuel savings will exceed financed cost.

IN THE **10th** YEAR — COSTS RETURNED

Blank form for this step on page 108

Though the Browns had hoped that the worksheet would show savings overtaking costs sooner, they expect their high-quality system to save increasingly greater amounts on gas bills well beyond its 10th year. They decide to proceed with their solar purchase.

TABLE E. Federal Energy Tax Credits	
If Your Installed Solar Hot Water System Cost Is: $	Then Your Tax Credit Is: $
1900	760
2050	820
2300	920
2550	1020

Full table on page 118

From TABLE E: The allowable Federal income tax credit is $820. The state energy office has told the Browns that they are entitled to an additional 25% state income tax credit of $513. Deduction of the total allowable credit of $1333

$2050 − $1333 = $717

yields a net system cost of $717.

TABLE J. Short-Term Loan Payment Factors

Term of Loan (Years)	Interest Rate (%)					
	8	9	10	11	12	13
1	1.04	1.05	1.06	1.06	1.07	1.07
2	1.09	1.10	1.11	1.12	1.13	1.14
3	1.14	1.14	1.16	1.18	1.20	1.21
4	1.17	1.19	1.22	1.24	1.26	1.29
5	1.22	1.25	1.28	1.31	1.34	1.37

Full table on page 121

From TABLE J: Multiplying by the loan payment factor corresponding to their expected 5-year, 13% home-improvement loan,

$717 × 1.31 = $982

the Browns find their overall financed cost of $982. When they divide this cost by first-year fuel bill savings (from Step 1) of $61,

$982 ÷ $61 = 16

a loan cost-to-savings ratio results.

TABLE H. Year In Which Accumulated Savings Exceed Total Costs

Cost/Savings Ratio	Fuel Escalation Rate (% per year)						
	4	6	8	10	12	14	16
10.	9	9	8	8	7	7	7
15.	12	12	11	10	10	9	9
20.	15	14	13	12	11	11	10
25.	18	16	15	14	13	12	11

Full table on page 120

From TABLE H: Using this ratio, the Browns learn that initial costs will be returned in the 10th year.

Part Two

A Buyer's Guide

Chapter 4

Choosing the Right Dealer/ Installer

In shopping for a solar water heater, the first choice to be made is not so much between *systems* as between *approaches*. Do you choose a system?—or do you seek a well-qualified dealer or installing contractor to guide your choice or make it for you?

Often there is no clear distinction between the two approaches. For example, system and dealer/ installer usually come in the same package when you buy from a major department store chain or on the basis of a well-known brand name.

In most cases, the wisest thing is to seek an *experienced* supplier who provides *everything,* installs it, and guarantees it. This could be a local plumbing or heating contractor. However, such individuals or firms are not available everywhere, and some purchasers will need to make their own ar-

rangements for installation. In fact, because the solar equipment industry is still relatively small, you may need to spend a considerable amount of time and effort on finding the right qualified, reliable dealer and installer.

Locating Solar Companies

Begin the search by asking your present heating dealer if he sells solar equipment or can recommend a solar supplier. The next step might be to:

Check the Yellow Pages under Solar Equipment.

Contact the National Solar Heating and Cooling Information Center at the address or toll-free number on the back cover of this guide. The Center can provide a list of "solar professionals"—architects, engineers, contractors, installers—who are active in your area. (Note that the appearance of names on a Center list does not constitute a recommendaton or endorsement.)

See if your state energy office or the nearest of the Department of Energy's four Regional Solar Energy Centers can put you in touch with installers or suggest other contacts. The local chapter of the American Section of the International Solar Energy Society may also provide some leads. (For help in locating these offices and centers, see pages 87-89.)

Call your local electric or gas utility. Its list of suppliers and contractors who sell and install energy-conservation measures in your locality may include solar dealer/installers.

Check solar catalogs and directories (several are listed on pages 92-93) at your local library.

Basic Qualifications

You will probably be consulting more than one dealer or contractor before selecting either a system or the individual or company who will install it. At the end of this section is a check list summarizing a number of characteristics—

more fully described below—that should be considered very definite assets in a solar-equipment or solar-contracting business. Look over the list, and decide which of these qualifications will be important considerations, perhaps even determining factors, in your choice of a dealer and/or installer:

Solar installation experience. Keep in mind that the best recommendation is having done the same job before—and well. Even suppliers who do no installation should have some expertise in installation methods and problems, including the site and building issues identified on page 22. Resist any suggestion that you contract installation to a local company with no solar energy system experience.

Satisfied customers. Ask for the names and addresses of a few recent buyers. Really satisfied customers are usually willing (and sometimes eager) to answer a few questions. How is the unit working? Has the dealer responded promptly to their questions and problems?

Solar-related background. There is a valuable carry-over from years of work in a field closely related to solar, such as mechanical engineering, plumbing, and heating and air-conditioning. Solar contractors with that kind of experience will have developed judgment in sizing heating systems according to demand and in balancing components. They may already be familiar with mineral-hardness or pH problems in the local water supply that could affect your choice of system type.

Nearby location. A local dealer or contractor can respond more quickly to problems. He has a local reputation to protect, and his previous installations probably are near enough for you to inspect.

Stable, established business. A well-established dealer with regular business hours can be counted on to provide promised estimates, keep appointments, and generally meet his obligations. He may even be in a position to help you get financing for your purchase.

Good complaint response record. Find out from the Better Business Bureau if it has received complaints about the dealer or contractor, and if action was taken to resolve any disputes. The various consumer, alternative energy, and solar energy organizations in your area also may have some knowledge of the company.

Strong manufacturer support. You can worry less about being one of the contractor's early customers if he has had the specialized training that prudent manufacturers provide to companies installing their systems. In some cases a manufacturer's representative will even inspect the finished installation.

Willingness to negotiate terms. There is every chance that you will want to modify the purchase/installation contract offered you. (See "Contracts," page 65.) Look for reasonable flexibility on this score.

Good warranty terms offered. At a minimum, installers should offer a year's full warranty on the installation as well as on all components and parts. Look for one who will make repairs on site or cover the cost of removing a defective component and shipping it for repair to a factory. These costs can be substantial. (For further discussion, see "Warranties," page 63.)

Professional standing. Not all jurisdictions require that solar energy systems be installed by a master plumber or licensed contractor, but it makes good sense to deal with a licensed, bonded craftsman. Groups like the National Association of Solar Contractors (910-17th Street, N.W., Suite 928, Washington, D.C. 20006) are trying to insure a level of quality in installations.

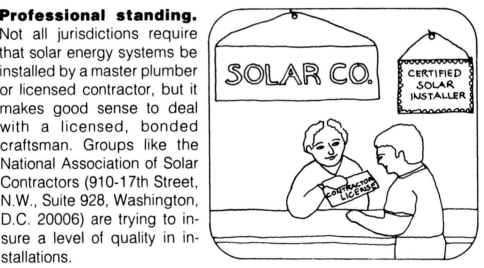

Familiarity with building regulations. Codes and building regulations are slowly beginning to change as states and communities try to adapt to new technologies. Your installer should be informed about current local regulations, particularly with respect to code and permit requirements.

Willingness to give full explanations. Make it plain that you will want a briefing on basic system operation and maintenance, in addition to a complete owner's information package. (See "Getting to Know the System," page 74.) The time the installer spends reviewing the system with you can save him unnecessary service calls.

Ongoing service offered. The difference between a good system and a poor one could come down to the availability of a competent mechanic to make a few needed adjustments. Will the installer be available to maintain and repair the system after the warranty period is over? Are frequently used parts kept in stock? Does he offer a full service contract? If he provides only warranted repairs and adjustments, your post-installation briefing on operating procedures and maintenance takes on added importance.

Checklist for Comparing
Solar Dealers/Installers

QUALIFICATIONS	Individuals or Firms		
	A	B	C
	(√) indicates description applies to individual or firm		
Solar installation experience			
Satisfied customers			
Solar-related background			
Nearby store or office			
Stable, established business			
Regular business hours			
Assistance in obtaining financing			
Good complaint response record			
Strong manufacturer support			
Willingness to negotiate terms			
Good warranty terms offered			
On-site repairs made			
Professional standing			
Familiarity with building regulations			
Willingness to give full explanations			
Ongoing service offered			
Frequently used parts in stock			

Note: Checklist for Comparing Solar Hot Water Systems appears on page 61.

Chapter 5

Choosing the Right System

The marketplace offers dozens of brand-name, packaged solar water heating systems, as well as a variety of separate components that can be hooked together to heat water effectively in different combinations. There is no "best" system. The right system for any buyer is quite simply the one that most closely matches his or her needs at an acceptable price.

In most cases, an experienced dealer or installer is the person most qualified to appraise particular needs and match them to the benefits of a particular system. This is so because the evaluations and trade-offs to be made are relatively complex:

Which type of system is right for you? Start by choosing between thermosiphoning and pumped, one-tank and two-tank, open-loop and closed-loop.

Which system will deliver the most heat per dollar of installed cost? Given the choice between two units of like quality, you will probably want to select the one that produces most for the money. Since system ratings are not yet available (see "Performance predictions," page 80), your comparisons will have to be based largely on dealer or manufacturer claims. Once again, the integrity and reputation of the individual or company proves to be central.

What about "high-efficiency" collectors? A collector is said to have "high efficiency" when it delivers more heat at low outside temperatures. For example, a double-glazed, flat-plate collector with a selective absorber surface has a higher efficiency than a single-glazed collector with a non-selective surface. In general, collector price rises with efficiency levels, and the more efficient collectors usually are justified only in systems used in severe climate areas. In any case, the collector should be evaluated as an integral part of the system. (See "Notes on System Efficiency," page 79.)

Do you keep your present water heater? In adding solar to a house, the useful life left in the existing water heater needs evaluation. Should it become the backup heater, fed by a new solar preheating tank? Or should it be replaced with a new unit matched to a solar tank and using the cheapest available fuel?

For many buyers, the best solution will be to rely on the recommendations of a qualified dealer and/or installing contractor. The information and guidelines on the following pages will help you to understand and evaluate recommendations.

Factors to Consider

In choosing a brand and model of packaged system or assembly of components, you will be weighing the total installed *cost* of various systems against each one's probable *quality*. When comparing two apparently very similar systems on the basis of price, be sure that they are alike in all specifics (same type, same size and capacity, comparable components).

Probably no single unit can offer all desired features at an affordable cost. Trade-offs will almost certainly have to be made. Some of the factors that should enter into your choice are:

System efficiency. How well does the unit heat water? (For discussion, see "Notes on System Efficiency," page 79.)

System durability. How long can it be expected to last before needing replacement? (See page 83.)

Manufacturers' commitment. A quality product is usually linked to good business practices. (See page 60.)

Warranty. If the system breaks down, will you be protected against unreasonable costs? (See page 63.)

Proven workability. Have these same components already been combined in a successful system? The only thing riskier than allowing a supplier to experiment with component sizing at your expense would be for you to create your own package from different sources.

Working systems. A chance to see the same model in operation and speak with the owners can be very instructive.

Maintenance. How much effort and expense will be needed to keep the system operating well?

Owner's information package. For good maintenance, it is vital that you be given full, clear, written information pertaining to your particular installation. (See page 74.)

Monitoring devices. Give preference to a system that offers, at least as options, a few gauges such as thermometers and a pump operating light to help you tell how well the sun is heating your water. (See page 13.)

Safety. As with any heating system, consider potential hazards and built-in safety precautions. (See page 17.)

A checklist at the end of this chapter (page 61) summarizes these and other good points to look for in a solar water heating system.

Product Literature

One way to get some idea of which system might be right for you is to explore free illustrated product literature from different manufacturers. If this is not readily available from dealers in your area, you can:

Write manufacturers. The local library may have one of the solar catalogs or directories (several of which are listed on page 92) that give the names and addresses of solar equipment manufacturers. Or, write or call the National Solar Information Center for a list of manufacturers whose products are distributed in your region.

Return reader service cards. Look for these in solar magazines in the library.

Watch for energy exhibits, heating-and-cooling industry trade shows, and the like. Solar product literature should be available at various booths.

Unfamiliar terms

Sometimes your best efforts to distinguish between competing systems on the basis of manufacturer information can be hampered by unfamiliar terminology and jargon. Terms and concepts that crop up regularly in solar water heating are defined or explained in the glossary beginning on page 95. Several of these are discussed at length in the appendices on system efficiency (page 79) and system durability (page 83.)

Manufacturer's Commitment

Certain solar hot water systems enjoy distinct advantages deriving from the manufacturer's demonstrated commitment to good service. Experience suggests that consumers should give preference to companies which:

Take responsibility for complete systems;

Train installation subcontractors;

Provide clear installation instructions;

Stand behind the product and its installation (see "Warranties," page 63);

Pre-package much of the plumbing and wiring;

Tell customers how to maintain their systems;

Offer, as part of the package, inexpensive thermometers and indicator lights needed to check whether the system is working well (see page 13);

Support installers in making on-site repairs.

When you buy from a reputable manufacturer, the possibility of unpleasant surprises during the lifetime of your system is reduced.

Checklist for Comparing Solar Hot Water Systems

	Systems		
	A	**B**	**C**
COST—including installation			
	(✓) indicates description applies to system		
BENEFITS **Efficiency**			
Proven workability			
Collector test results available			
Working system available to inspect			
Durability			
Well-made collectors			
Reputable brands			
Low maintenance requirements			
Easily accessible components			
Good protection against freezing			
Good protection against corrosion			
Good protection against leaks			
Built-in safety precautions			
Manufacturer's Commitment			
Installer training provided			
Replacement parts stocked			
Repairs made on site or costs covered			
Good warranty offered through dealer			
Complete owner's information provided			
Monitoring Devices			
Thermometers and gauges included			
Pump operating light included			
Devices available as options			

Note: Checklist for Comparing Solar Dealers/Installers appears on page 55

Warranties and Contracts

The same two written statements of understanding apply to the usual solar purchase as to any other major consumer purchase. These are *a warranty* and *a contract*.

Warranties

Warranties are the express commitment of the manufacturer and installer to stand behind the product and its installation and to protect the purchaser from unreasonable costs due to defects in materials and workmanship. Very few solar equipment warranties address levels of performance, since conditions not under the manufacturer's control can affect operating efficiency.

In this area, too, solar shoppers have to evaluate risks and make trade-offs. Those who choose a

simple, low-cost system may be satisfied with limited warranty coverage or even none. Others will be willing to pay the premium for generous, no-risk coverage, or will settle for something in between. Each individual consumer has to decide how high a value to place on guaranteed protection against unanticipated expense due to system failure. (Service contracts, another form of protection, are discussed in the next chapter.)

The logical time to add warranty provisions to the delicate balancing of factors that goes into choosing the right system is after the field has been narrowed to perhaps one or two good systems suiting the consumer's identified needs. This is so because the warranty is only as sound as the product it covers and the company behind it.

What the consumer really wants is a good *system*, not a "good" *warranty* covering an inferior system which, if defective by its very nature, can never be brought up to par. Worse yet are overgenerous promises stemming from the inexperience of a manufacturer who has no mechanism for handling complaints and perhaps only limited expectations of staying in business for the duration of the promises.

However, it is not unreasonable to expect a good system to be covered by a good warranty. Industry and consumer spokesmen generally agree that solar buyers should receive full backing from both manufacturer and installer at least during the shakedown period when most defects show up. Owners of conventional heating systems customarily receive this full parts-and-labor coverage for at least one year. Among solar companies, there is a growing trend to cover major components—such as collectors, heat exchanger, and storage tank—for as long as three to five years (or to provide a service contract).

Warranty limitations

Be alert to limitations on coverage in limited warranties. *Some of these can have costly consequences.* For example, under some warranties an owner can be forced to bear the cost of removing a defective component, shipping it to a distant factory, and reinstalling it once repaired. Other warranties cover these costs as well as provide for on-site repairs.

Pass-through warranties

A solar energy system is usually a composite of various manufacturers' parts. To get all the parts warranted, some buyers have had to accept several separate warranties. Quite a cast of characters (including several manufacturers, a packager, a dealer, and an installer) can become involved in the debate over who is responsible for a specific repair.

The best way out of this dilemma, which reflects the newness of the industry, is for manufacturers and packagers to warrant their equipment to the buyer through the installing contractor, who is likely to be the person or firm making the repairs. Then the installer's warranty not only covers defects resulting from improper installation but also includes all the "pass-through" equipment warranties. This approach, which is gaining in popularity, protects the installer when an installation defect is actually caused by inadequate manufacturer instructions.

In your own case, try to determine any distinctions between the manufacturer's and the contractor's responsibilities to you. Clarify this with your contractor.

Contracts

As with any home-improvement project, carefully read the contract offered to you for signature. Whether a document is termed an estimate, a bid, or a contract is of little consequence as long as your signature commits you to purchase the system. Everything promised should be spelled out in writing at a price that is clearly *firm*.

The typical home-improvement contract includes such items as a detailed description of the work to be completed, the total contract price, approximate dates for the beginning and completion of work, and a general description of the warranty to be provided, in addition to various customary certifications about liability, liens, and so on.

In the case of your water heater installation, you may want to add a limitation on the permissible interruption of hot water service—one or two days should be sufficient—and a stipulation that a percentage of the contract price will be withheld pending the installer's re-inspection of the system after a few weeks' normal operation. It also makes sense to spell out the installer's responsibility to restore disturbed roofing, siding, building insulation, walls, and landscaping following completion of installation.

Seek legal assistance if you have any problems negotiating terms with your contractor or in writing them into the contract.

Consumer protection measures.

A federal regulation permitting three days in which to reconsider a signed contract applies to you if nearly all of your conversations with the sales representative or contractor took place away from his or her place of business, or were by telephone. If you wish to cancel the contract, you must notify the contractor in writing within the three-day period.

Even more stringent consumer protection measures may apply in your state. For information, contact the nearest state or local office of consumer affairs or consumer protection agency. (For help with consumer complaints, see page 77.)

Part Three

Protecting Your Investment

Chapter 7

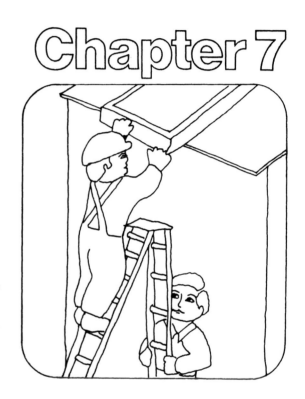

Getting a Good Installation

The best way to insure that your newly purchased water heating system actually returns the fuel savings that you are counting on is to hire a reliable installer. If you have followed the recommendations of earlier chapters, you have selected a well qualified individual or firm, and have chosen an appropriate solar hot water system with his or her help. You can trust a skilled installer to install the unit correctly, in accordance with the manufacturer's instructions.

Since you are quite likely to become a "sidewalk superintendent" watching the process, these general pointers* will help you understand what is going on, and permit you to ask about anything that seems questionable:

*For more detailed information, obtain a copy of HUD's *Installation Guidelines for Solar DHW Systems*. This report, which is based on experiences with thousands of solar hot water installations, may be of particular interest to your solar contractor; ordering information appears in the Introduction footnote.

Insulation. Because heat losses can occur at many points in a solar energy system, all piping, ducts, tanks, and heat exchangers must be heavily insulated. Three inches of fiberglass around the storage tank is by no means excessive. Insulation used outdoors needs to be waterproof and ultraviolet-resistant.

Collector location. Collector panels are best located as near as possible to the storage tank to minimize heat loss and the need for a larger pump. Collectors should not be placed where snow and ice could build up at their base—blocking sunlight or forming an ice dam—or slide off onto walks or entrances.

Roof mountings. Collectors are usually mounted almost flush against the roof unless a pronounced deviation from correct tilt justifies the added cost of a support rack. Spacers are used to raise the collectors slightly, preventing damage to the roof from moisture collection, except where, as in new construction, the collectors can be mounted directly on the roofing felt, and flashed and sealed as part of the weather surface. Holes where pipes enter the roof must be carefully flashed and sealed to prevent leaks and heat loss.

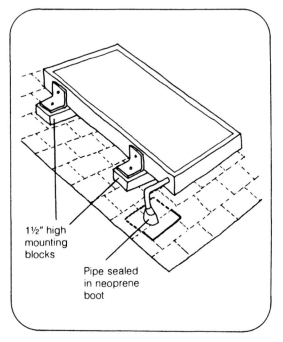

1½" high mounting blocks

Pipe sealed in neoprene boot

Typical roof mounting for existing house.

Ground mountings. The lower edge of ground-mounted collectors should be located high enough above the ground to reduce mud splashing and coverage by drifting snow; this distance will usually be at least 18 inches.

Storage tank. Outlets should be located so that spills or overflows will not damage the premises.

Pitch of pipes. If the system drains down, there must be no traps or horizontal runs in outside pipes where water could collect and freeze.

Ease of repair. In good collector installations, cracked or warped glazing can be replaced without dismantling the entire array of panels. Providing some working space around components will make it easier to check liquid levels and lubricate pumps and fans. Controls, valves, joints, and other potential trouble areas should be accessible.

Mixing valve. If tank temperature is likely to get hot enough to burn someone's hand (more than 140°F), a valve is needed to mix in some cold water before delivery to the tap.

Pressure testing. Testing to detect leaks must be done before insulation is applied to the pipes, in order to avoid possible water damage to the insulation.

Plumbing inspection. The local plumbing inspector is primarily concerned with safety violations, and usually will schedule his inspection while the pipes are still uncovered.

Living with Your Solar Energy System

Chapter 8

Once a solar unit is installed, it profits from regular maintenance. If you have not already discussed ongoing maintenance with your installer, by all means do so now. It may be that he will guarantee his work only so long as you see to it that certain maintenance is provided. Perhaps he offers this under a service contract similar to those available for other heating equipment. In many cases, service contracts prove to be a good value.

This chapter offers some general pointers on maintenance, and also suggests some ways that the owner's participation can help keep a system working as it should. The following points are discussed:

To use the system effectively, you need to understand how it operates. When you

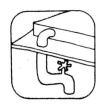

know what is supposed to happen, you are more alert to situations which are not normal, but serve as warning flags.

Occasionally repairs need to be made. If defects are caught early, the repairs usually are simple and inexpensive.

If the system breaks down and you feel that the dealer/installer or manufacturer is not responsive, there are several sources of help. (See "Consumer Complaints," page 77.)

Getting to Know the System

Operating test

When your solar water heater is finally in place, the installer should test it thoroughly and then start it up. Try to be on hand for the operating test, so that you can observe both filling and draining procedures. The time that the installer spends reviewing the system with you may be to his or her benefit as well as yours, since an informed owner can often detect problems before they become really troublesome.

As you look over the system:

Watch for signs that any equipment has not been well installed, such as loose tape at joints or gaps in insulation.

Locate any gauges, thermometers, or signal devices that indicate how the system is working.

Find out how to adjust the temperature settings, and observe the controls and sensors in operation.

Ask how the freeze protection mechanism works. Has it been tested?

Note the location of shutoff valves and switches, and of the fuse or circuit breaker. Are they properly labeled? Labeling showing flow direction and operating temperatures and pressures can be helpful.

If yours is a closed-loop system, check to see that the fill valve on the collector loop has been tagged to indicate when it was filled and with what fluid.

Ask if potential health or safety hazards are clearly flagged. These might include points where toxic fluid or fluid under high temperature or pressure could be discharged.

Owner's information package

Whether your system comes with a formal owner's manual or with a packet of separate pieces makes little difference. What is highly important is that the information provided be both *clear* and *thorough*. Before the installer leaves, review the information package with him or her. Be sure it contains:

Instructions on starting and shutting down the system.

A basic parts list.

A maintenance schedule listing all items requiring periodic attention.

Diagrams of your particular installation showing valves, wiring, switches, sensors, controls, and connections to the backup system.

Procedures for vacation shutdown, if this is recommended, and for emergencies.

Identification of any safety or health hazard.

Warranty information.

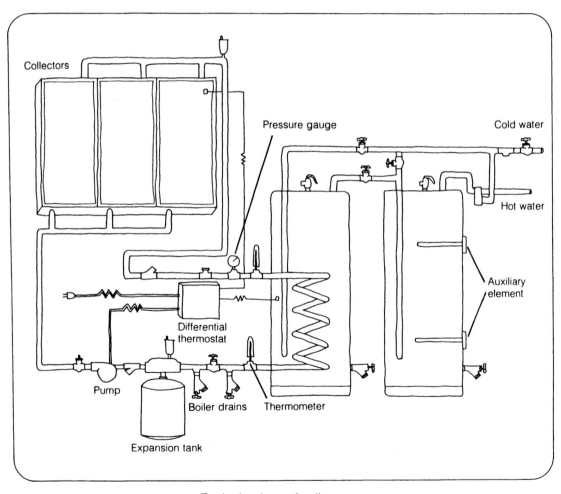

Typical schematic diagram

Shakedown period

Consider the first weeks of operation as a shakedown period for your new solar water heater. Relatively simple adjustments are often needed before a system will perform as well as expected. For this reason, the installer should probably return to check it over after a few weeks have elapsed.

Note that savings will be greatest if you avoid placing heavy demands on your hot water supply over a short period. Try to schedule most washing activities during hours when the sun is shining.

Routine maintenance

If you plan to provide your own routine maintenance, take a close look at the manufacturer's recommended maintenance schedule. Make some notations on your calendar if you are likely to need reminding. Two components that call for regular attention are:

Heat Transfer fluid. In a closed loop system, check the liquid level every few months and refill as needed, after checking for leaks. Test and/or replace the fluid at the manufacturer's suggested intervals. (Antifreeze solutions can break down into highly corrosive liquid.)

Collectors. Keep the glazing free of leaves and other debris, and wash off dust and dirt each spring and fall. At the same time, inspect the collector for signs of deterioration or damage.

Signs of trouble

If the hot water taps are delivering scalding hot water or no hot water at all, everyone will notice that something is wrong with the water heating system. These less obvious signs also give warning of probable trouble:

The pump runs all the time or at night. By operating on a sunless day or at night, it can release heat from storage to the atmosphere.

The pump is off on sunny days. Unless there is plenty of hot water on tap, the pump should be operating to move heated fluid from the collectors to storage.

The solar storage tank fails to heat up after the pump has been running on sunny days.

Tank temperature drops quickly even though little or no hot water is being used.

The backup heater turns on when the storage tank is hot.

The reading on the collector-loop pressure gauge drops significantly.

Fluid is dripping from a pipe or tank.

Report defects of this kind to your dealer or installer as soon as they are noticed.

A simple test

Perhaps the simplest way to find out whether your solar unit is delivering hot water is to turn off the backup heater. Choose a clear, sunny morning when there are several loads of laundry to be done or some other practical way of drawing off the hot water already in storage. Late in the day, water from a hot water tap should be at least warm. If not, the solar unit may need adjustment or repair.

To cut off your backup electric heater temporarily, trip the circuit breaker or remove the fuse that protects the heater from overload. At the end of the test, flip the circuit breaker back into place or replace the fuse.

To reduce the fuel supplied to your gas heater, turn the control to its lowest (pilot light) setting. (Caution: do not turn *off* the pilot light.)

Consumer Complaints

Dealers and installers are usually eager to protect their reputations for reliability and good service. Most defects in equipment or installation can be worked out to everyone's satisfaction. Occasionally, however, an individual or firm is unresponsive to a serious complaint or unfairly evades responsibility.

When that happens, make a note of pertinent information that has been transmitted over the telephone or in person, and record the date. These records, and copies of any correspondence between you and the company, will be invaluable if a prolonged dispute should develop.

As with any consumer complaint, there are several avenues that a system owner can follow. Depending on the seriousness of the grievance, you may choose to:

File a complaint with the local Better Business Bureau. The Bureau will press the company for some resolution of the problem, and its response becomes a matter of record.

Write the manufacturer. A reputable firm will make some effort to resolve the problem through the distributor, if one is involved, or will take some direct action.

Contact the state consumer protection office. (This may be a separate agency or part of the State Attorney General's office. See page 89.) Many states have consumer protection laws that go way beyond the level of Federal protection. The consumer protection staff will provide

information and advice, and may even be willing to mediate your dispute.

Contact consumer complaint centers at the county or municipality level or in the private sector.

Take the matter to Small Claims Court. This can be done without an attorney. However, the small-claims dollar limit or ceiling may rule out your action if more than a few hundred dollars are involved.

Although the Federal Trade Commission does not intervene in *individual* consumer complaints, it is continually looking for evidence of misleading sales practices that could be nationwide or commonplace in an industry. If you choose to inform them about your grievance, write to: Energy & Product Information Division, Bureau of Consumer Protection, Federal Trade Commission, Washington, D.C. 20580. If a breach of warranty is involved, write instead to: Public Reference Branch, Federal Trade Commission, Washington, DC 20580, or call (202) 523-3598.

Summary

Although solar domestic hot water systems have been used for decades in Florida and California, they are new to most parts of the country. Being new, and therefore unfamiliar, they tend to seem overly complicated to many people. In this report, we have attempted to "uncomplicate" solar energy to help you understand what a solar system can do for you, and to define factors that need to be considered in deciding to install a solar system.

A solar hot water system is an expensive consumer product, and you should exercise all of the cautions you would use for any other large purchase:

1. Select the seller and installer carefully, based on their reputations, experience, and recommendations of others.

2. Be sure you understand the purchase order or contract before you sign it, and do not sign until you agree to the terms.

3. Obtain clear operating and maintenance instructions, and assure yourself that you can obtain repairs and service (both during and after the warranty period).

4. Become familiar with the system as installed so you know what to look for in verifying its performance and in identifying problems before they become serious.

We believe that a solar hot water system will save energy and money for many people, and by following the recommendations of this report, you will enjoy many years of satisfactory performance. We are glad you are interested in solar energy.

Appendix A.
Notes on System Efficiency

A consumer shopping for a solar energy system is likely to see and hear a lot about "thermal performance" or "efficiency." Product literature often claims a certain degree of efficiency for a collector. This is simply a reference to how well the collector can do its job, which is to capture the sun's heat energy and pass it along.

Collector efficiency is a measure of the energy absorbed by the absorber plate, and of the rate at which heat is removed to storage or is lost from the collector. The amount of available sunshine fluctuates widely depending on geographical location, weather, and time of day. Environmental conditions such as wind have a major effect. A single-glazed collector (one covered by a single pane of glass) that works at 65 percent efficiency under clear summer skies (meaning that it is capable of capturing 65 percent of the solar energy striking its surface under those conditions) will obviously collect less energy when the sky is hazy or cloudy. In addition, as outside temperatures drop, collector efficiency falls off due to heat loss through the cover plate.

Efficiency myths

Three misconceptions are fairly common:

Myth #1: *Efficient collectors guarantee good system performance.* Not true. The overall efficiency of a solar water heating system depends on many factors in addition to collector performance. Both the quality of system installation and the degree to which all components of the system are correctly sized and matched to each other are of paramount importance. Heat losses from uninsulated pipes or a poorly insulated hot water storage tank will undercut efficiency disastrously. Inadequate storage capacity or heat exchanger surface has a similarly negative effect.

Myth #2: *High-efficiency collectors are always the best buy.* Not true. A collector that gathers more heat than is needed will probably release it back into the atmosphere. You will want to pay only for the level of efficiency that saves you money. A solar dealer or other qualified professional who has assessed your household's use of hot water can help you get the most energy for your money, up to the total required for your needs but not much beyond.

Myth #3: *High collector temperature means high efficiency*. Not true. Heat losses increase as the collector temperature rises above that of surrounding air. Flat-plate collectors actually deliver more heat energy when operating at relatively low temperatures—between 100° and 150°F.

Performance predictions

With the aid of a computer, an experienced professional can estimate the amount of heat that a particular collector will deliver per square foot of collecting surface under given conditions, and expand this to an estimate of the total energy that a specific system might supply in a particular application over a year's time.

However, these data have distinct drawbacks:

Predictions of system performance are only that—predictions. Actual *measured* performance away from a laboratory often falls far below the predicted level.

Manufacturers sometimes base efficiency claims on a maximum figure rather than the far more significant yearly average.

What are really needed are *comparative* ratings to indicate a system's *average* heat output over an extended period. Fortunately for the consumer, government and industry are now working together to develop such ratings.

Efficiency ratings

Collector efficiency ratings are the first step. Two states, Florida and California, now test and rate solar collectors. In addition, two industry groups—the Solar Energy Industries Association (SEIA) and the Air Conditioning and Refrigeration Institute (ARI) are developing a joint program to certify collector performance.

This type of chart is often found in product literature. It represents one collector's performance when tested according to ASHRAE Standard 93-77, a widely accepted testing method. From the chart a qualified professional can determine the rate at which the collector's efficiency decreases, and this can be a crucial factor in trade-offs between collectors. Average efficiency over the year will be considerably below the percentage shown at the top of the curve. (See Table K, page 122, for a method of comparing performance charts.)

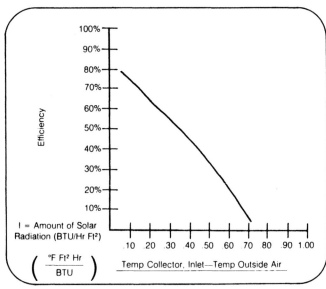

The American Society of Heating, Refrigerating and Air Conditioning Engineers (ASHRAE) is taking the process a step further with its work on development of a test procedure for *full* solar hot water *systems*. It will become much easier to compare solar units when system ratings are available. In the meantime, look closely at BTU-output claims for evidence that they reflect yearly averages for your own region.

Appendix B.
Notes on System Durability

Solar water heaters justify their purchase cost only if they save on utility bills over a long period. A good system can be expected to last 15 to 20 years with correct maintenance. However, good workmanship and the right choice of materials are major factors in durability. Only an experienced person can examine a component or system and juggle all the trade-offs that may be involved in the choice of materials and manufacturing processes. Consult your contractor or seek other expert advice when sizing up the quality and merits of a particular unit.

Well-made collectors

Collectors are subjected to blazing sunlight, intense thermal expansion and contraction, pressure, corrosive forces, and both extremes of temperature. The soundness of design and workmanship obviously bears directly not only on lifespan but also on performance. If a panel is poorly constructed, it will not collect the needed heat, no matter how carefully installed.

If you have the opportunity to see a system in operation, you can often tell something about how well the collectors are withstanding the elements. Signs of poor weathering include:

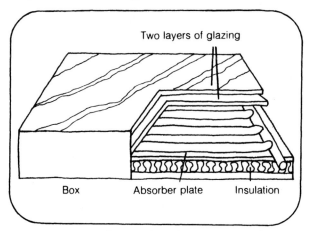

A well-made collector has insulation underneath and along the sides of the absorber plate. Its covering of glass or a long-lasting plastic may be either one or two layers, depending on the efficiency required. The enclosure will be weatherproof and generally airtight.

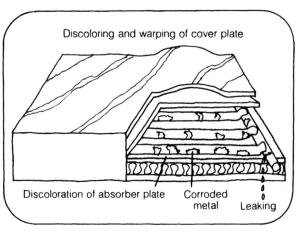

Signs of poor weathering

Leaks;

Cracks, warping, or discoloration in the transparent cover plate;

Dirt streaks, condensation, or a film on the inner surface of the cover plate;

Fading or discoloration of the absorber plate;

Broken seals, torn or displaced gaskets, or visible dust inside the cover plate;

Peeling coatings, pitting of metal parts, or rust.

Reputable brands

With the notable exception of its collectors, a solar water heating system is made up mostly of parts available from any plumbing supply outlet. (See the full system diagram on page 75.) But not all pumps, pipes, tanks, valves, and controls are of lasting quality. Ask for proven brands or look for packaged systems from reputable manufacturers.

Weakening Forces

A well-designed solar hot water system has internal safeguards against freezing, corrosion, and leaks. As you look over a system, ask the dealer or installer to explain what provisions have been made for protection against these three major weakening forces. Typical provisions are described below.

Protection against freezing

Four methods are commonly used to prevent collectors from freezing:

Draindown. A fail-safe valve causes water to be dumped from the collectors whenever the pump stops running. (See illustration, page 15.)

Drainback. Collector fluid drains into a holding tank when the pump stops running. (See illustration, page 16.)

Antifreeze. The heat transfer fluid contains antifreeze.

Applied heat. At near-freezing temperatures, the collectors are bathed automatically with warm water pumped from storage or are warmed by attached electric heat tapes. Neither process can take place during a power failure, which is a major drawback. Moreover, the cost of drawing off heat from storage and of operating electric warmers makes applied heat impractical except in locations where freezing temperatures are rare.

Protection against corrosion

Metal pitting can irreparably shorten a water heater's life span. Of the metals most commonly used in collectors, only copper will last a long time without special protection against the minerals in water. However, less expensive aluminum and steel stand up well if corrosion-inhibiting chemicals are added to the water and maintained at full strength.

Another type of corrosion occurs when any two dissimilar metals, such as copper and steel, are in direct contact in moisture. Such a "galvanic" reaction can be avoided or kept to a minimum by good system design and proper selection of components.

Protection against leaks

Leaks from or within a collector or elsewhere in the system are less likely if a special solder for high temperatures is used on internal joints. Also, the system should have gaskets and seals selected to perform at high temperatures.

Appendix C.
Solar Information Resources

1. Organizations

A. Government-Sponsored Organizations

National Solar Energy Centers

THE NATIONAL SOLAR HEATING AND COOLING INFORMATION CENTER

    P.O. Box 1607
    Rockville, Maryland 20850
    Toll-free numbers
    800/523-2929; in Pennsylvania,   800/462-4983
                in Alaska or Hawaii,   800/523-4700

The National Solar Information Center collects, organizes, and distributes information on solar technology in an effort to encourage the energy-saving use of solar heating and cooling in homes and other buildings. The Center is operated by the Franklin Research Center for the U.S. Department of Housing and Urban Development in cooperation with the U.S. Department of Energy.

For complete information on resources and services available to the consumer through the Center, or to order any of the free publications listed on page 90, return the information request form on the last page of this guide.

SOLAR ENERGY RESEARCH INSTITUTE (SERI)

    1617 Cole Boulevard
    Golden, Colorado 80401
    303/231-1000

The Solar Energy Research Institute, which supports the solar research, development, and demonstration activities of the U.S. Department of Energy, is developing specialized solar energy data bases. Its Solar Energy Information Data Bank (SEIDB) answers technical inquiries and conducts computerized information searches.

Regional Solar Energy Centers

The four regional centers sponsored by the U.S. Department of Energy are collecting and distributing solar information particularly suited to the specific region covered. Center locations and the states served by each are given on the following page.

MID-AMERICAN SOLAR ENERGY COMPLEX (MASEC)

8140 26 Street
Bloomington, Minnesota 55420
612/853-0400

Serves Illinois, Indiana, Iowa, Kansas, Michigan, Minnesota, Missouri, Nebraska, North Dakota, Ohio, South Dakota, and Wisconsin.

NORTHEAST SOLAR ENERGY CENTER (NESEC)

470 Atlantic Ave.
Boston, Massachusetts 02110
617/292-9250

Serves Connecticut, Maine, Massachusetts, New Hampshire, New Jersey, New York, Pennsylvania, Rhode Island, and Vermont.

WESTERN SUN

715 S.W. Morrison
Portland, Oregon 97205
503/241-1222

Serves Alaska, Arizona, California, Colorado, Hawaii, Idaho, Montana, Nevada, New Mexico, Oregon, Utah, Washington, and Wyoming.

SOUTHERN SOLAR ENERGY CENTER

61 Perimeter Park
Atlanta, Georgia 30341
404/458-8765

Serves Alabama, Arkansas, Delaware, District of Columbia, Florida, Georgia, Kentucky, Louisiana, Maryland, Mississippi, North Carolina, Oklahoma, Puerto Rico, South Carolina, Tennessee, Texas, Virginia, West Virginia, and the Virgin Islands.

Energy Extension Service

The U.S. Department of Energy operates an Energy Extension Service in all 50 states. Under the EES grant program, each state decides which small-scale energy users need information and assistance, what types of services they need, and what institutions in the state can best deliver those services. Typical activities include energy audits, workshops, and energy hotlines. Check with your state energy office for further information.

State Energy Offices

Every state has an office that responds to public requests for energy-related information. To locate the energy office in your state, call the National Solar Heating and Cooling Information Center at its toll-free number. Or, use the form at the back of this guide to request the Center's *list of solar information resources in your state*.

State Consumer Protection Offices

The state consumer protection office is the best source of information on the consumer protection laws that apply in a particular state; these measures often go way beyond the level of Federal protection. If you need help locating the office in your state (it may be a separate agency or part of the State Attorney General's office), call the National Solar Heating and Cooling Information Center at its toll-free number.

B. Privately Sponsored Organizations

AMERICAN SECTION OF THE
INTERNATIONAL SOLAR ENERGY SOCIETY, INC.
American Technological University
P.O. Box 1416
Killeen, Texas 76541
817/526-1300

The American Section of ISES provides an interdisciplinary forum for groups and individuals concerned with the use of solar energy. Its 27 chapters promote education and distribute information on regional, state, and local levels. Meetings are usually open to the public and offer an opportunity for interested lay people to talk with experienced professionals and other solar enthusiasts.

To locate the chapter nearest you, contact the national office at the address above.

SOLAR ENERGY INDUSTRIES ASSOCIATION (SEIA)
1001 Connecticut Avenue NW, Suite 800
Washington, D.C. 20036
202/293-2981

SEIA is a trade association representing many manufacturers and distributors of solar equipment. Its official monthly publication—"Solar Engineering"—and its annual catalog/solar industry index are among the publications listed on page 92.

SHEET METAL AND AIR CONDITIONING CONTRACTORS'
NATIONAL ASSOCIATION (SMACNA)
8224 Old Court House Road
Vienna, Virginia 22180
703/790-9890

This trade association will refer consumers to local contractors experienced in installation of air-type solar heating systems. SMACNA's home-study course, "Fundamentals of Solar Heating,"is listed among the publications on page 91.

2. Publications

A. Lists from the National Solar Heating and Cooling Information Center

The National Solar Information Center publishes a *separate* list for *each* of the 50 states in *each* of the following categories:

Manufacturers. Solar equipment manufacturers located within the state, with an indication of the type(s) of products manufactured.

Solar Professionals. Individuals and firms providing solar-related services (design, energy consultation, engineering, installation, construction) in the state.

Solar Buildings. Private residences and other solar-equipped buildings in the state that may be visited, either with or without appointment.

Solar Legislation. Brief descriptions of state legislation on taxation, grants and loans, land use, standards, and building codes, with contacts for further information.

Colleges and Universities with Solar Courses. Schools offering solar-related courses, with a brief description of the courses.

Local Solar Information Resources. How to contact the state energy office; the Regional Solar Energy Center serving the state; the Energy Extension Service office, if any; local chapters of solar energy societies and certain other local groups. Also, articles, reports, and books relating to use of solar energy in the state.

To request the lists that apply to your state, return the form on the last page of this guide.

B. Consumer Guides

The Connecticut Solar Handbook. Connecticut Citizen Action Group, Box G, Hartford, CT 06106: 1978, 47 pp., $.75.
    Helpful tips on choosing a solar energy system and dealer/installer.

Solar for Your Present Home. California Energy Commission, 1111 Howe Avenue, Sacramento, CA 95825: 1978, 162 pp., $4.00.
    Full information on types of solar energy systems; worksheets covering every step in retrofitting an existing house with a space and/or water heating unit.

The Buy Wise Guide to Solar Heat by F. Hickok. Hour House, P.O. Box 40082, St. Petersburg, FL 33743: 1976, 121 pp., $9.00.
    What to be aware of when shopping for solar equipment.

Homeowner's Guide to Solar Heating and Cooling by W. M. Foster. Tab Books, Blue Ridge Summit, PA 17124: 1976, 196 pp., $4.95.
    How to purchase, install and maintain solar heating and cooling and hot water systems.

Home Guide to Solar Heating and Cooling by I. Hand. Harper and Row, New York, NY 1977, 183 pp., $3.95.

How to Buy Solar Heating . . . Without Getting Burnt! by M. Wells and I. Spetgang. Rodale Press, Emmaus, PA 18049: 1978, 262 pp., $6.95.
> Focuses mainly on solar space heating, but offers useful consumer tips.

Solar Decision Book: Your Guide to Making a Sound Investment by R. H. Montgomery and J. Budnick. Dow Corning Corporation, Midland, MI 48640: 1978, 266 pp., $10.00.
> Very complete information geared to decisions to be made at every stage of planning a solar energy system.

Sunset Homeowner's Guide to Solar Heating, Sunset Books, Lane Publishing, Menlo Park, CA: 1978, 96pp., $2.95.

C. Do-It-Yourself

Build Your Own Solar Water Heater. Florida Conservation Foundation, Inc., 935 Orange Ave., Winter Park, FL 32789: 1976, 25 pp., $2.50.
> Directions for building a thermosiphoning or a simple pumped system.

The Solar Home Book by B. Anderson with M. Riordan. Brick House Publishing Co., 3 Main St., Andover, MA 01810: 1976, 297 pp., $7.50.
> Focuses mainly on low-cost applications of solar space heating, but do-it-yourself water heating is covered.

Installation Guidelines for Solar Domestic Hot Water Systems in One- and Two-Family Dwellings. U.S. Department of Housing and Urban Development; Superintendent of Documents, U.S. Government Printing Office, Washington, DC 20402, Stock No. 023-000-00520-4: 1979, 111 pp., $4.00.
> Useful supplement to manufacturer's instructions for the skilled homeowner and the professional installation contractor.

Build Your Own Solar Water Heater by S. Campbell with D. Taff. Garden Way Publishing, Charlotte, VT 05445: 1978, 109 pp., $7.95.
> Comprehensive treatment aimed at the non-technical reader. Fully illustrated.

Solar Water Heating for the Handyman by S. Paige. Edmund Scientific Company, 101 E. Gloucester Pike, Barrington, NJ 08007: 1974, 32 pp., $4.00.
> Descriptions of solar water heaters from simple to complex, sizing guidelines, and three brief sample heater plans.

Fundamentals of Solar Heating. Sheet Metal and Air Conditioning Contractors' National Association (SMACNA) and Northamerican Heating and Air-Conditioning Wholesalers Association (NHAW); National Technical Information Service, 5285 Port Royal Road, Springfield, VA 22131, Stock No. HCP/M4038-01: 1978, 188 pp., $4.50. Offered as accredited home-study course by the Home Study Institute, 1661 West Henderson Road, Columbus, OH 43220: $125.00; $95.50 for SMACNA or NHAW member.
> Training course in sizing, installing, and servicing solar heating systems. Separate lesson on water heating systems.

D. Technical Books

Basics of Solar Heating and Hot Water Systems. American Institute of Architects/ Research Corporation, 1735 New York Ave., N.W., Washington, DC 20006: 1977, 48 pp., $5.00.

System design considerations for both passive and active approaches to home heating and hot water systems.

Design Manual for Solar Water Heaters. Horizon Industries, 12606 Burton St., North Hollywood, CA 91605: 1977, 40 pp., $5.00.

Manual introducing the engineer to various systems.

E. Catalogs and Directories

The catalogs and directories listed here provide information on solar energy products and/or solar professionals and organizations. Unless the publication's title indicates a regional focus, coverage is nation-wide. State-level solar directories are published and distributed free by several of the state energy offices.

SEM-79 (Solar Engineering Master Catalog and Solar Industry Index). Solar Engineering Publishers, Inc., 8435 N. Stemmons Freeway, Dallas, TX 75247: 1978, 224 pp., $15.00.

Extensive product listings, with specifications. Manufacturers are cross-referenced to products and trademarks. List of solar professionals.

Solar Age Resource Book. Solar Age, Church Hill, Harrisville, NH 03450: 1979, 242 pp., $9.95.

Buyers' guide for all types of solar products, with a geographical directory of architectural and design services. Articles on related subjects.

The 1979 Sun Catalog. Solar Usage Now, Inc., Box 306, Bascom, OH 44809: 1979, 287 pp., $2.00.

Kits and components for do-it-yourselfers.

Directory of the Solar Industry. Solar Data, 13 Evergreen Rd., Hampton, NH 03842: 1978, $20.00.

Solar companies listed alphabetically and by state.

Edmund Scientific Catalog. Edmund Scientific Company, 300 Edscorp Building. Barrington, NJ 08007: 1979, $1.00

Quite a few solar products.

New England "Yellow Pages" of Solar Energy Development. New England Solar Energy Association, P.O. Box 541, Brattleboro, VT 05301: 1978, 70 pp., $3.00.

Solar products and services, and alternative energy resources.

Western Regional Solar Energy Directory. Southern California Solar Energy Association, 202 C Street–1A, San Diego, CA 92101: 1978, 67 pp., $2.35.

Informal Directory of the Organizations and People Involved in the Solar Heating of Buildings by W. A. Shurcliff. 19 Appleton St., Cambridge, MA 02138: 1977, 3rd edition, $9.00 prepaid.

F. Periodicals

Solar Age, Church Hill, Harrisville, NH 03450. $20.00/year.
Official magazine of the American Section of the International Solar Energy Society, Inc.

Solar Engineering, 8435 Stemmons Freeway, Suite 880, Dallas, TX 75247. $15.00/year.
Official publication of the Solar Energy Industries Association.

Appendix D.
Glossary

Terms Used in Solar Water Heating

Absorber plate
Black surface inside a collector that absorbs solar radiation and converts it to heat.

Active system
Solar heating system that, unlike a "passive" system, requires external mechanical power to move the collected heat.

Air-type system
Active system that uses air as the medium to collect and carry solar heat; not commonly used for water heating alone because of the greater cost and poorer efficiency of air-to-water heat exchangers.

Ambient temperature
Temperature of the surroundings: for collectors, outdoor temperature; for storage, that of the space in which the tank is located.

ASHRAE
American Society of Heating, Refrigerating and Air-Conditioning Engineers; a professional organization that has developed standards for testing.

Auxiliary heat
See Backup system.

Backup system
Conventionally fueled (i.e., with gas, oil, or electricity) unit capable of meeting the hot water demand when the solar energy system is not operating or producing enough.

Block rates
See Rate schedule.

British thermal unit (BTU)
Heat needed to raise 1 pound (1 pint) of water 1°F. BTU are used to express the amount of energy used in heating.

Cash flow analysis

Comparison of solar energy costs and savings on a month-by-month basis. Positive (or favorable) cash flow begins when monthly fuel bill savings exceed monthly solar loan payments.

Closed-loop (or indirect) system

Active system in which the sun's heat passes to household water indirectly. Solar energy is collected by a fluid such as antifreeze solution or a special oil that is repeatedly recirculated through a closed pipe loop. Heat from this fluid is transferred to the household water through the wall(s) of a heat exchanger at the storage tank.

Collector

Device that absorbs solar radiation and converts it to heat. *See* Concentrating collector; Flat-plate collector.

Collector coolant

See Heat-transfer fluid.

Collector efficiency

See Efficiency.

Collector orientation

See Orientation.

Collector tilt (or slope)

Angle at which a collector is raised from the horizontal in order best to capture the sun's rays.

Concentrating collector

Collector with lenses or reflectors that concentrate the sun's rays on a relatively small absorber surface; most concentrators sacrifice efficiency compared to flat-plate collectors in order to obtain higher temperature.

Conduction

Flow of heat from a warmer material to a cooler one in direct physical contact.

Convection

Heat transfer through moving currents of air or liquid. *Natural* convection occurs when the fall by gravity of heavier cool fluid causes a lighter warm fluid to rise. (*See* Thermosiphoning.) *Forced* convection occurs when the fluid is circulated by a fan or pump.

Corrosion

Deterioration of metal by chemical action. *See also* Galvanic corrosion.

Cover plate

Glass or transparent plastic placed over the flat-plate collector to trap heat.

Diffuse radiation

Solar radiation that has been scattered by clouds and particles in the atmosphere and casts no shadow; flat-plate collectors can absorb it, but concentrating collectors cannot.

Direct radiation

Solar radiation that comes straight from the sun, casting shadows on a clear day.

Direct system

See Open-loop system.

Double-glazed

Covered by two panes of glass or other transparent material.

Double-wall heat exchanger

One that provides two distinct walls between the heat-transfer fluid and the household water; such a separation is necessary when the heat-transfer fluid contains non-drinkable substances.

Drainback

Applies to a closed-loop water heating system in which the heat transfer fluid drains into a holding tank when the pump is turned off, thereby protecting the collectors from freezing.

Draindown

Applies to an open-loop water heating system in which water from the collectors is dumped into a sump or drain when the pump is turned off, thereby preventing freeze damage.

Efficiency

Ratio of the heat delivered by a system to the solar energy supplied to it. Efficiency can be measured with respect to the system as a whole, or the focus can be on a particular component's performance. For example, when half of the incoming sunlight is passed along as heat, a *collector* is said to be 50 percent efficient; if half of that heat is then lost from the pipes and storage tank, *system* efficiency would be 25 percent.

Expansion tank

In a closed-loop system, a tank that provides space for the expanded volume of the heat-transfer liquid as it is heated in the pressurized collector loop.

Flat-plate collector (or collector panel)

Solar collection device in which sunlight striking a flat black surface is converted to heat without being concentrated.

Forced convection

See Convection.

Fossil fuels

Fuels such as oil, natural gas, and coal, formed from organic material deposited millions of years ago.

Galvanic corrosion

Deterioration of metallic parts occurring when a heat-transfer fluid that conducts electricity (water and antifreeze solution do; silicone and hydrocarbon oils do not) contacts two dissimilar metals that are not properly isolated electrically.

Glazing

Transparent or translucent covering (e.g., glass or plastic) that reduces heat loss from a solar collector panel.

Heat exchanger

A metal surface, such as a coiled copper tube immersed in a tank of water, that transfers heat from one fluid to another. *See also* Double-wall heat exchanger.

Heat transfer

Movement of heat from one place to another by conduction, convection, or radiation.

Heat-transfer fluid (or medium)

Fluid that carries heat from solar collector to storage. In direct systems, this is plain water; in indirect systems, it may be treated water, air, or a special oil. Since it reduces collector temperature by carrying away heat, this fluid is sometimes referred to as "collector coolant."

Hydronic

A heating system using circulating hot water.

IMPS

The U.S. Department of Housing and Urban Development's "Intermediate Minimum Property Standards Supplement: Solar Heating and Domestic Hot Water Systems" is sometimes referred to by this acronym. To be in compliance with IMPS, all parts of a system must be constructed and installed with the minimum level of quality required by the standards.

Indirect system

See Closed-loop system.

Insolation

Total amount of solar radiation (direct, diffuse, and reflected) striking a collector cover plate; usually measured in BTU per square foot per hour or day.

Insulation

Material with high resistance (R-value) to heat flow.

Internal rate of return

Expected rate of return if a solar water heater is viewed as an investment.

Kilowatt-hour (kWh)

Unit of electricity equal to 3,413 BTU; domestic water heating for a family of four requires 400 kWh per month on the average.

Life-cycle cost

Total cost to buy and run a solar water heating system over a period of years; in the typical analysis, this cost is compared to like costs of a conventionally fueled system.

Liquid-type system

One that uses liquid rather than air as the medium to collect and carry solar heat.

Mixing (or tempering) valve

One that adds cold water to the hot water being drawn from storage as needed to maintain a preset safe temperature.

Natural convection

See Convection.

One-tank system

Solar water heating system using a single tank to store all household hot water, whether heated in the solar collectors or by an auxiliary electric unit.

Open-loop (or direct) system

Solar water heating system in which household water is heated directly in the solar collectors. Circulation may be forced by a pump or it may be by natural means (thermosiphoning and gravity flow).

Orientation

Number of degrees to the east or west of south that a collection surface faces.

Passive system

With reference to water heating, a thermosiphoning system. *See* Thermosiphoning.

Payback period

Time needed for fuel bill savings to return the initial cost of a solar unit, plus any interest.

Peak load

Maximum demand for electrical power experienced during a defined period by a utility company. Since peak load determines the total generating capacity required, some utilities have begun to encourage conservation by charging highest prices during peak demand periods.

Potable

Suitable for drinking or cooking purposes.

Preheating

Use of a solar energy system to raise the temperature of water supplied to a conventional water heater.

Radiation

Flow of energy from one body to another across a space. Radiation may proceed via visible light or via invisible infrared rays, depending on the temperature of the radiating body. *See also* Solar radiation.

Rate schedule

Rules determining the cost of electricity or gas used during a billing period. Typically, the first group or block of kilowatt-hours or of therms (or hundreds of cubic feet of gas) costs more than the second block; the second more than the third, and so on.

Reflected radiation

Sunlight that is reflected from terrain or buildings.

Reradiation

Emission of previously absorbed radiation via invisible infrared heat.

Retrofit

Applies to renovations that are done to save energy; to retrofit with solar means to equip an existing building with a solar energy system.

R-value

See Thermal resistance.

Selective surface

Absorber-plate coating that absorbs visible light, but cannot emit infrared radiation, thereby reducing heat losses.

Sensors

See Temperature sensors.

Single-glazed

Covered by one pane of glass or other transparent material.

Sizing

Determining the size or capacity of separate components in a solar water heating system (especially collector area and storage tank capacity) that will provide the greatest benefit per dollar of investment.

Solar access

Access to direct sunlight without interference from vegetation or structures on neighboring property. Such access is generally not protected by law; private agreements can sometimes be worked out among neighbors.

Solar collector

See Collector.

Solar fraction

Percentage of heating requirement provided by a solar energy system.

Solar noon

Time of day when the sun reaches its maximum altitude for the day and is due south; this occurs exactly midway between sunrise and sunset.

Solar radiation

Energy from the sun which comes to earth in the form of direct, diffuse, or reflected rays.

Stagnation

High-temperature state that occurs when the sun is shining on a collector but no fluid is flowing through to carry heat away; can be damaging to the collector.

Storage

Container and medium that absorb collected solar heat and hold it for later use.

Stratification

In the context of water heating systems, the tendency of stored water to remain in layers of different temperature, with the coldest on the bottom and the warmest on top.

Temperature sensors

Devices that measure the temperature of the collectors and storage and send signals to the control unit to start or stop pumps or actuate valves.

Tempering valve

See Mixing valve.

Therm

A quantity of natural gas, approximately 100 cubic feet, that provides 100,000 BTU when burned.

Thermal performance

See Efficiency.

Thermal resistance (R-value)

Tendency of a material to retard the flow of heat; the higher the R-value, the greater the insulating value of the material.

Thermosiphoning

Upward movement, by natural convection, of fluid warmed in a solar collector to a storage tank directly above it.

Tilt angle

See Collector tilt.

Transfer fluid

See Heat-transfer fluid.

Two-tank system

Solar water heating system using a separate tank holding solar-heated water as a preheater for a smaller, conventionally fueled water heater. Two tanks must be used when the backup fuel is gas or oil.

U-value

Rate of heat flow by conduction through a complete building section, such as a wall; the lower the U-value, the greater the insulating value of that section.

Appendix E.
Economic Worksheets
for Solar Water Heating

These step-by-step worksheets are designed to help the consumer evaluate the costs and savings of solar water heating. Simply by completing *two* of the four steps, you can estimate how much solar energy would cut your annual fuel bills and how long it should take for those fuel bill savings to exceed the system's cost or, in the case of mortgages, improve your cash flow.

Before working through the two steps that apply to your situation, look up or identify the following information:

collector area suited to your hot water use (can be estimated from TABLE C, page 117);

cost of the installed system (can be estimated from TABLE D, page 118);

estimated financing terms (number of years, interest rate), unless you will be paying cash; and

current unit price of the fuel now heating your water. (For fuel oil, look up the cost per gallon on your latest bill. For gas or electricity, ask the utility company to tell you how much was charged for the *last* therm or kilowatt-hour that you used in the latest billing period—this "marginal" cost should be used in preference to any averaged figure.)

You may also want to look back at the filled-in sample worksheets that appear on pages 36-47 of this guide. Three fully worked examples illustrate use of the tables and the simple computations required.

If your solar water heater would be:

covered by a **mortgage** financing a new home, complete only

STEP 1 Estimate Fuel Bill Savings	STEP 2 Compare Costs And Savings (Mortgages)

See Example 1, pp. 36-39

purchased with **cash,** complete only

| STEP 1
Estimate
Fuel Bill
Savings | STEP 3
Compare Costs
And Savings
(Cash Purchases) |

See Example 2, pp. 40-43

purchased through a **home-improvement loan,** complete only

| STEP 1
Estimate
Fuel Bill
Savings | STEP 4
Compare Costs
And Savings
(Short-Term Loans) |

See Example 3, pp. 44-47

If you guess at system costs in order to complete STEP 2, 3, or 4, take another look at the figures later when you have a contractor's estimate. You may want to refigure using the firm price.

To compare two or more solar water heaters, photocopy the worksheets before filling in the blanks.

STEP 1. Estimate Fuel Bill Savings

Procedure

1a. In TABLE A (pages 110-115), find the number of units of electricity (page 110), natural gas (page 112), or oil (page 114) that your solar water heater would save annually. Fill in here.

ANNUAL ENERGY SAVINGS

1b. Fill in the current Unit Price of that fuel, and MULTIPLY.   × $

UNIT PRICE OF FUEL

You can expect a solar water heater to cut the first year's fuel bills about this much.   = $

FIRST-YEAR FUEL BILL SAVINGS

OPTIONAL

1c. To estimate fuel savings over a longer period, refer to TABLE B (page 116). Find the cumulative savings factor for the period of years that you are considering. Fill in here, and MULTIPLY.   ×

CUMULATIVE SAVINGS FACTOR

Over that time period, a solar water heater can cut your fuel bills about this much.   = $

LONG-TERM FUEL BILL SAVINGS

Factors to Consider

Actual energy savings may vary somewhat from the amount estimated, depending on such factors as climate and shading at your site and the degree to which conservation measures are employed.

Since fuel prices seem likely to rise steeply in the near term, actual first-year fuel bill savings could turn out to be 15 to 20 percent greater than the amount estimated, which is based on current fuel prices.

Savings over a longer period are tied to the annual rate of fuel price escalation, which is very difficult to predict over the long term.

If you plan to:

finance as part of a mortgage on a new home—**Go on to STEP 2**

pay cash—**Go on to STEP 3**

finance with a short-term loan—**Go on to STEP 4**

STEP 2. Compare Costs and Savings—
Mortgage Loans

Procedure

2a. From TABLE D (page 118), or from your dealer's estimate, fill in your Installed System Cost.

$ [   ] **INSTALLED SYSTEM COST**

2b. From TABLE E (page 118), fill in the Federal Energy Tax Credit on this cost, and SUBTRACT.

− $ [   ] **ENERGY TAX CREDIT**

This is what your system will cost you after you claim the Federal tax credit. (See note below on state tax credits.)

= $ [   ] **NET SYSTEM COST**

2c. From TABLE F (page 119), select the annual loan payment factor corresponding to your mortgage's probable interest rate and term. Fill in here, and MULTIPLY.

× [   ] **LOAN PAYMENT FACTOR**

This is the portion of your annual mortgage payment that covers your solar water heating system.

= $ [   ] **ANNUAL SOLAR LOAN COST**

2d. From STEP 1, fill in First-Year Fuel Bill Savings, and DIVIDE.

÷ $ [   ] **FIRST-YEAR FUEL BILL SAVINGS**

This is the ratio of your annual solar loan cost to first-year savings.

= [   ] **LOAN COST/ SAVINGS RATIO**

2e. From TABLE G (page 119), fill in the year during which annual fuel savings will exceed annual loan costs and create a favorable cash flow.

IN THE [   ] YEAR **FAVORABLE CASH FLOW BEGINS**

Factors to Consider

State tax credits, where available, reduce net cost still further. For details, contact your state energy office.

Much of the initial solar cost may be recovered when a house is sold. To the extent that this happens, fuel bill savings are a bonus. Solar-equipped houses will probably sell more easily and quickly than other houses as the fuel situation gets worse.

There are real advantages to having a solar energy system built *into* a new house rather than added on later. The risk of installaton defects is less if the system is assembled with the roof and walls. Costs are lower as well, and can be included in the mortgage, meaning that monthly payments are kept low.

STEP 3. Compare Costs and Savings— Cash Purchases

Procedure

3a. From TABLE D (page 118), or from your dealer's estimate, fill in your Installed System Cost.

$	**INSTALLED SYSTEM COST**

3b. From TABLE E (page 118), fill in the Federal Energy Tax Credit on this cost, and SUBTRACT.

—	$	**ENERGY TAX CREDIT**

This is what your system will cost you after you claim the Federal Tax Credit. (See note below on state tax credits.)

=	$	**NET SYSTEM COST**

3c. From STEP 1, fill in First-Year Fuel Bill Savings, and DIVIDE.

÷	$	**FIRST-YEAR FUEL BILL SAVINGS**

This is the ratio of your system's net cost to first-year savings.

=		**COST SAVINGS RATIO**

3d. From Table H (page 120), fill in the year during which total fuel savings will exceed initial costs.

IN THE _____ YEAR	**COSTS RETURNED**

Factors to Consider

State tax credits, where available, reduce net cost still further. For details, contact your state energy office.

Much of the initial solar cost may be recovered when a house is sold. To the extent that this happens, fuel bill savings are a bonus. Solar-equipped houses will probably sell more easily and quickly than other houses as the fuel situation gets worse.

The cash you invest in a solar unit is likely to earn a better return in *tax-free* savings than a savings account would pay in *taxable* interest (see page 33). Though some years may pass before total fuel bill savings equal your initial outlay, your monthly utility bill will be cut from the outset, so improved cash flow begins at once.

STEP 4. Compare Costs and Savings— Short-Term Loans

Procedure

4a. From TABLE D (page 118), or from your dealer's estimate, fill in your Installed System Cost.

$ ____ **INSTALLED SYSTEM COST**

4b. From TABLE E (page 118), fill in the Federal Energy Tax Credit on this cost, and SUBTRACT.

− $ ____ **ENERGY TAX CREDIT**

This is what your system will cost you after you claim the Federal Tax Credit. (See note below on state tax credits.)

= $ ____ **NET SYSTEM COST**

4c. From TABLE J (page 121), select the loan payment factor corresponding to your loan term and interest rate, and MULTIPLY.

× ____ **SHORT-TERM LOAN FACTOR**

This is the total amount you will pay out over the life of the loan, including interest.

= $ ____ **OVERALL FINANCED COST**

4d. From STEP 1, fill in First-Year Fuel Bill Savings, and DIVIDE.

÷ $ ____ **FIRST-YEAR FUEL BILL SAVINGS**

This is the ratio of your overall financed cost to first-year savings.

= ____ **LOAN COST/ SAVINGS RATIO**

4e. From TABLE H (page 120), fill in the year during which total fuel savings will exceed financed cost.

IN THE ____ YEAR **COSTS RETURNED**

Factors to Consider

State tax credits, where available, reduce net cost still further. For details, contact your state energy office.

Much of the initial solar cost may be recovered when a house is sold. To the extent that this happens, fuel bill savings are a bonus. Solar-equipped houses will probably sell more easily and quickly than other houses as the fuel situation gets worse.

Because the short duration of home-improvement loans makes monthly payments relatively high, a solar unit will usually improve cash flow only after the loan is repaid. However, savings keep growing as long as the system lasts—15 or 20 years, if it is well made.

A Note On Assumptions

The tables on the following pages were prepared from *averaged* data furnished by RSVP, the U.S. Department of Housing and Urban Development's computer program for estimating solar economics. Annual energy savings (TABLE A) and suggested collector sizes (TABLE C) are based on:

Weather data for a *typical* year at each of the cities listed. Varying conditions from year to year will affect the fuel actually saved.

Assumed daily use of 20 gallons of water heated to 140°F by the first two occupants of a house, with each additional person using 15 gallons. Fuel savings will be greater if conservation measures are taken, such as setting the hot water thermostat at 120° to 130°F.

Average performance of a *typical* system—deemed for this purpose to be an open loop, two-tank, draindown system with storage capacity of 1.8 gallons for each square foot of collector area. Assumed collector performance (performance curve slope of .83 BTU/hr./sq. ft/degree F; y-intercept of .73) resembles that provided by flat steel absorber plates with a selective black coating under one layer of glazing. However, some collectors on the market may vary as much as 20% from this average performance. Annual Energy Savings figures can be adjusted to reflect data on the performance of a specific collector when this data is available (see TABLE K, page 122.)

The *average* operating efficiency of conventional water heaters according to fuel used—specifically, electricity, 100%; natural gas, 60%; and fuel oil, 50%.

Maintenance and any additional property tax or insurance charges due to a solar water heating system have not been included in costs. These tend to be offset by Federal tax deductions for interest payments.

TABLE A. Annual Energy Savings From A Typical Solar Water Heater

1. Kilowatt-Hours Of Electricity Saved

Note: To learn the collector area best suited to the *average* household of your size in your location, refer to Table C, page 117. Figures there and in this table are based on *averaged* data described in "A Note on Assumptions," page 109.

Number of Occupants	2		4			6 (or more)			
Collector Area (sq. ft.)	40	60	40	60	80	40	60	80	100
Location									
ALABAMA									
BIRMINGHAM	1900	2300	2300	3100	3600	2500	3400	4200	4800
ALASKA									
FAIRBANKS	2000	2600	2300	3200	3900	2500	3500	4300	5100
ARIZONA									
TUCSON	2500	2600	3300	4100	4400	3700	4900	5700	6100
ARKANSAS									
LITTLE ROCK	2000	2400	2500	3300	3800	2700	3700	4500	5100
CALIFORNIA									
LOS ANGELES	2300	2700	2900	3800	4300	3200	4300	5200	5900
SACRAMENTO	2300	2500	3000	3800	4100	3300	4400	5200	5700
SAN FRANCISCO	2300	2700	2900	3800	4400	3200	4300	5200	5900
COLORADO									
DENVER	2900	3100	3600	4600	5200	3900	5300	6400	7200
GRAND JUNCTION	2700	2900	3400	4400	4900	3700	5000	6000	6700
CONNECTICUT									
HARTFORD	1600	2100	1900	2600	3200	2000	2800	3500	4100
DELAWARE									
WILMINGTON	1900	2300	2200	3000	3500	2300	3200	4000	4700
DISTRICT OF COLUMBIA									
WASHINGTON	1800	2200	2100	2900	3400	2300	3100	3900	4500
FLORIDA									
JACKSONVILLE	1900	2300	2400	3100	3600	2600	3600	4300	4900
MIAMI	1800	2100	2300	3000	3400	2600	3400	4100	4600
TALLAHASSEE	2000	2300	2400	3200	3600	2600	3600	4300	4900
TAMPA	1900	2200	2500	3200	3600	2700	3600	4300	4900
GEORGIA									
ATLANTA	1900	2400	2300	3100	3700	2500	3500	4200	4900
SAVANNAH	2100	2400	2600	3300	3800	2800	3800	4600	5200
HAWAII									
HILO	1800	2100	2300	3000	3400	2500	3400	4100	4600
HONOLULU	2200	2300	3000	3700	3900	3400	4400	5100	5500
IDAHO									
BOISE	2400	2700	3000	3900	4400	3200	4400	5300	6000
POCATELLO	2500	2900	3100	4100	4700	3400	4600	5600	6400
ILLINOIS									
CHICAGO	1900	2400	2200	3000	3600	2400	3300	4100	4800
PEORIA	2100	2600	2500	3400	4100	2700	3800	4700	5400
INDIANA									
INDIANAPOLIS	1800	2200	2000	2800	3300	2200	3000	3800	4400
IOWA									
DES MOINES	2200	2600	2600	3400	4100	2700	3800	4700	5400
KANSAS									
WICHITA	2300	2700	2900	3800	4400	3100	4300	5200	5900
KENTUCKY									
LEXINGTON	1800	2300	2100	2900	3400	2300	3200	3900	4500
LOUISVILLE	1800	2300	2100	2900	3500	2300	3200	3900	4600
LOUISIANA									
NEW ORLEANS	1900	2300	2400	3100	3600	2600	3500	4300	4800
SHREVEPORT	2000	2300	2500	3200	3700	2700	3700	4400	5000
MAINE									
CARIBOU	1900	2400	2100	2900	3600	2200	3100	3900	4600
PORTLAND	1700	2200	1900	2700	3300	2000	2900	3600	4200
MARYLAND									
BALTIMORE	1900	2300	2200	2900	3500	2300	3200	4000	4700
MASSACHUSETTS									
AMHERST	1800	2300	2100	2900	3500	2200	3100	3900	4600
BOSTON	1700	2200	2000	2700	3300	2100	2900	3700	4300
MICHIGAN									
LANSING	2000	2500	2300	3100	3800	2500	3400	4300	5000
SAULT ST. MARIE	1700	2200	1900	2600	3200	2000	2800	3600	4200

Number of Occupants	2		4			6 (or more)			
Collector Area (sq. ft.)	40	60	40	60	80	40	60	80	100
Location									
MINNESOTA									
MINN-ST. PAUL	2000	2500	2300	3100	3800	2400	3400	4300	5000
MISSISSIPPI									
JACKSON	1900	2300	2400	3100	3600	2600	3500	4300	4900
MISSOURI									
KANSAS CITY	2100	2500	2500	3300	4000	2700	3700	4600	5300
ST. LOUIS	2000	2400	2400	3200	3800	2600	3600	4400	5100
MONTANA									
BILLINGS	2300	2700	2700	3700	4300	2900	4100	5000	5800
GREAT FALLS	2200	2700	2600	3500	4200	2800	3900	4800	5500
NEBRASKA									
LINCOLN	2300	2800	2800	3700	4400	3000	4100	5000	5800
NEVADA									
LAS VEGAS	2500	2600	3500	4200	4400	3900	5100	5900	6200
RENO	2900	3100	3700	4700	5200	4000	5400	6500	7200
NEW HAMPSHIRE									
CONCORD	1700	2200	1900	2600	3200	2000	2800	3600	4200
NEW JERSEY									
ATLANTIC CITY	2200	2600	2600	3500	4100	2800	3900	4700	5400
NEW MEXICO									
ALBUQUERQUE	2800	3000	3600	4600	5000	4000	5300	6400	7000
NEW YORK									
ALBANY	1900	2400	2300	3100	3700	2400	3400	4200	4900
NEW YORK	1600	2100	1900	2600	3100	2000	2800	3500	4100
ROCHESTER	1500	2000	1800	2400	2900	1800	2600	3300	3800
SYRACUSE	1500	2000	1700	2400	2900	1800	2600	3300	3800
NORTH CAROLINA									
CAPE HATTERAS	2000	2400	2400	3200	3700	2600	3600	4400	5000
RALEIGH	1900	2300	2300	3100	3600	2500	3400	4200	4800
NORTH DAKOTA									
BISMARCK	2300	2800	2700	3600	4300	2800	3900	4900	5700
OHIO									
CLEVELAND	1600	2000	1800	2500	3000	1900	2700	3400	4000
COLUMBUS	1700	2100	1900	2600	3200	2000	2900	3600	4200
OKLAHOMA									
OKLAHOMA CITY	2200	2600	2700	3500	4100	2900	4000	4800	5500
TULSA	2000	2400	2500	3300	3800	2700	3700	4500	5100
OREGON									
MEDFORD	2000	2300	2400	3200	3700	2600	3600	4400	5000
PORTLAND	1500	1900	1800	2400	2900	1900	2700	3300	3900
PENNSYLVANIA									
PHILADELPHIA	1800	2300	2100	2800	3400	2200	3100	3900	4500
PITTSBURGH	2000	2500	2400	3200	3900	2600	3600	4400	5100
STATE COLLEGE	1900	2400	2200	3000	3600	2300	3300	4100	4700
RHODE ISLAND									
PROVIDENCE	1800	2200	2000	2800	3400	2100	3000	3700	4400
SOUTH CAROLINA									
CHARLESTON	1900	2300	2300	3000	3500	2500	3400	4100	4700
SOUTH DAKOTA									
RAPID CITY	2300	2800	2800	3700	4400	3000	4100	5100	5800
TENNESSEE									
NASHVILLE	1800	2200	2200	2900	3400	2300	3200	3900	4500
TEXAS									
AMARILLO	2600	2800	3200	4100	4700	3500	4700	5700	6400
DALLAS	2000	2300	2500	3300	3800	2700	3700	4500	5100
EL PASO	2600	2700	3500	4300	4600	3900	5100	6000	6400
HOUSTON	1800	2100	2200	2800	3300	2300	3200	3900	4400
UTAH									
SALT LAKE CITY	2600	2800	3200	4200	4600	3500	4700	5700	6400
VERMONT									
BURLINGTON	1900	2400	2200	2900	3600	2300	3200	4000	4700
VIRGINIA									
MT. WEATHER	2100	2600	2500	3300	4000	2600	3700	4600	5300
NORFOLK	2000	2400	2400	3200	3800	2600	3500	4300	5000
RICHMOND	1900	2300	2200	3000	3600	2400	3300	4100	4700
WASHINGTON									
SEATTLE	1600	2000	1900	2600	3100	2100	2900	3500	4100
SPOKANE	2000	2400	2300	3100	3700	2500	3500	4300	4900
WEST VIRGINIA									
CHARLESTON	1600	2100	1900	2600	3100	2000	2800	3500	4100
WISCONSIN									
MADISON	2000	2500	2300	3100	3800	2400	3400	4300	5000
WYOMING									
CASPER	2800	3100	3400	4500	5100	3700	5000	6100	6900
CHEYENNE	2700	3100	3200	4300	5000	3500	4800	5800	6700

TABLE A. Annual Energy Savings From A Typical Solar Water Heater

2. Therms Of Natural Gas Saved

Note: To learn the collector area best suited to the *average* household of your size in your location, refer to Table C, page 117. Figures there and in this table are based on *averaged* data described in ''A Note on Assumptions,'' page 109.

Number of Occupants	2		4			6 (or more)			
Collector Area (sq. ft.)	40	60	40	60	80	40	60	80	100
Location									
ALABAMA									
BIRMINGHAM	110	130	130	170	200	140	190	240	270
ALASKA									
FAIRBANKS	120	150	130	180	220	140	200	250	290
ARIZONA									
TUCSON	140	150	190	230	250	210	280	330	350
ARKANSAS									
LITTLE ROCK	120	140	140	190	220	150	210	250	290
CALIFORNIA									
LOS ANGELES	130	150	170	220	250	180	250	300	330
SACRAMENTO	130	140	170	210	230	190	250	290	320
SAN FRANCISCO	130	150	170	220	250	180	250	300	330
COLORADO									
DENVER	160	180	200	260	300	220	300	360	410
GRAND JUNCTION	150	170	190	250	280	210	280	340	380
CONNECTICUT									
HARTFORD	90	120	110	150	180	110	160	200	230
DELAWARE									
WILMINGTON	110	130	120	170	200	130	180	230	270
DISTRICT OF COLUMBIA									
WASHINGTON	100	130	120	160	190	130	180	220	260
FLORIDA									
JACKSONVILLE	110	130	140	180	210	150	200	240	280
MIAMI	100	120	130	170	190	150	200	230	260
TALLAHASSEE	110	130	140	180	210	150	200	250	280
TAMPA	110	130	140	180	200	150	210	250	280
GEORGIA									
ATLANTA	110	130	130	180	210	140	200	240	280
SAVANNAH	120	140	150	190	220	160	220	260	290
HAWAII									
HILO	110	120	130	170	190	140	200	230	260
HONOLULU	120	130	170	210	220	190	250	290	310
IDAHO									
BOISE	140	150	170	220	250	180	250	300	340
POCATELLO	140	160	180	230	270	190	260	320	360
ILLINOIS									
CHICAGO	110	140	130	170	210	140	190	230	270
PEORIA	120	150	140	190	230	150	210	260	310
INDIANA									
INDIANAPOLIS	100	130	120	160	190	120	170	210	250
IOWA									
DES MOINES	120	150	150	200	230	160	220	270	310
KANSAS									
WICHITA	130	150	160	210	250	180	240	290	330
KENTUCKY									
LEXINGTON	100	130	120	160	200	130	180	220	260
LOUISVILLE	100	130	120	160	200	130	180	220	260
LOUISIANA									
NEW ORLEANS	110	130	140	180	200	150	200	240	270
SHREVEPORT	110	130	140	180	210	150	210	250	290
MAINE									
CARIBOU	110	140	120	170	200	130	180	220	260
PORTLAND	100	130	110	150	190	110	160	200	240
MARYLAND									
BALTIMORE	110	130	120	170	200	130	180	230	260
MASSACHUSETTS									
AMHERST	100	130	120	160	200	130	180	220	260
BOSTON	100	120	110	150	190	120	170	210	240
MICHIGAN									
LANSING	110	140	130	180	220	140	200	240	280
SAULT ST. MARIE	100	120	110	150	180	110	160	200	240

TABLE A-2 Continued from page 112

Continued from page 112

Number of Occupants	2		4			6 (or more)			
Collector Area (sq. ft.)	40	60	40	60	80	40	60	80	100
Location)									
MINNESOTA									
MINN-ST. PAUL	110	140	130	180	220	140	190	240	280
MISSISSIPPI									
JACKSON	110	130	140	180	210	150	200	240	280
MISSOURI									
KANSAS CITY	120	140	140	190	230	150	210	260	300
ST. LOUIS	120	140	140	180	220	150	200	250	290
MONTANA									
BILLINGS	130	160	160	210	250	170	230	290	330
GREAT FALLS	130	150	150	200	240	160	220	270	320
NEBRASKA									
LINCOLN	130	160	160	210	250	170	230	290	330
NEVADA									
LAS VEGAS	140	150	200	240	250	220	290	340	360
RENO	160	180	210	270	300	230	310	370	410
NEW HAMPSHIRE									
CONCORD	100	120	110	150	180	110	160	200	240
NEW JERSEY									
ATLANTIC CITY	120	150	150	200	230	160	220	270	310
NEW MEXICO									
ALBUQUERQUE	160	170	210	260	290	230	300	360	400
NEW YORK									
ALBANY	110	140	130	170	210	140	190	240	280
NEW YORK	90	120	110	150	180	110	160	200	230
ROCHESTER	90	110	100	140	170	110	150	190	220
SYRACUSE	90	110	100	140	170	100	150	190	220
NORTH CAROLINA									
CAPE HATTERAS	110	140	140	180	210	150	200	250	280
RALEIGH	110	130	130	170	210	140	190	240	270
NORTH DAKOTA									
BISMARCK	130	160	150	200	250	160	220	280	320
OHIO									
CLEVELAND	90	120	100	140	170	110	150	190	230
COLUMBUS	100	120	110	150	180	120	160	200	240
OKLAHOMA									
OKLAHOMA CITY	120	150	150	200	230	160	230	270	310
TULSA	120	140	140	190	220	150	210	250	290
OREGON									
MEDFORD	110	130	140	180	210	150	200	250	280
PORTLAND	90	110	100	140	170	110	150	190	220
PENNSYLVANIA									
PHILADELPHIA	100	130	120	160	190	130	180	220	260
PITTSBURGH	120	140	140	180	220	150	200	250	290
STATE COLLEGE	110	140	130	170	210	130	190	230	270
RHODE ISLAND									
PROVIDENCE	100	130	110	160	190	120	170	210	250
SOUTH CAROLINA									
CHARLESTON	110	130	130	170	200	140	190	230	270
SOUTH DAKOTA									
RAPID CITY	130	160	160	210	250	170	230	290	330
TENNESSEE									
NASHVILLE	100	130	120	160	200	130	180	220	260
TEXAS									
AMARILLO	150	160	180	240	270	200	270	320	360
DALLAS	120	130	140	190	210	160	210	250	290
EL PASO	150	150	200	250	260	220	290	340	370
HOUSTON	100	120	120	160	190	130	180	220	250
UTAH									
SALT LAKE CITY	150	160	180	240	260	200	270	330	360
VERMONT									
BURLINGTON	110	130	120	170	200	130	180	230	270
VIRGINIA									
MT. WEATHER	120	150	140	190	230	150	210	260	300
NORFOLK	110	140	140	180	210	150	200	250	280
RICHMOND	110	130	130	170	200	140	190	230	270
WASHINGTON									
SEATTLE	90	120	110	150	180	120	160	200	230
SPOKANE	110	130	130	180	210	140	200	240	280
WEST VIRGINIA									
CHARLESTON	90	120	110	150	180	110	160	200	230
WISCONSIN									
MADISON	110	140	130	180	220	140	190	240	280
WYOMING									
CASPER	160	180	190	250	290	210	290	350	390
CHEYENNE	150	180	180	240	280	200	270	330	380

TABLE A. Annual Energy Savings From A Typical Solar Water Heater

3. Gallons Of Fuel Oil Saved

Note: To learn the collector area best suited to the *average* household of your size in your location, refer to Table C, page 117. Figures there and in this table are based on *averaged* data described in "A Note on Assumptions," page 109.

Number of Occupants	2		4			6 (or more)			
Collector Area (sq. ft.)	40	60	40	60	80	40	60	80	100
Location									
ALABAMA									
BIRMINGHAM	90	110	110	150	180	120	170	200	230
ALASKA									
FAIRBANKS	100	130	110	160	190	120	170	210	250
ARIZONA									
TUCSON	120	120	160	200	210	180	240	280	300
ARKANSAS									
LITTLE ROCK	100	120	120	160	190	130	180	220	250
CALIFORNIA									
LOS ANGELES	110	130	140	180	210	160	210	250	290
SACRAMENTO	110	120	140	180	200	160	210	250	280
SAN FRANCISCO	110	130	140	180	210	150	210	250	290
COLORADO									
DENVER	140	150	180	230	250	190	260	310	350
GRAND JUNCTION	130	140	160	210	240	180	240	290	330
CONNECTICUT									
HARTFORD	80	100	90	130	150	100	140	170	200
DELAWARE									
WILMINGTON	90	110	110	140	170	110	160	200	230
DISTRICT OF COLUMBIA									
WASHINGTON	90	110	100	140	170	110	150	190	220
FLORIDA									
JACKSONVILLE	90	110	120	150	180	130	170	210	240
MIAMI	90	100	110	150	170	120	170	200	220
TALLAHASSEE	100	110	120	150	180	130	170	210	240
TAMPA	90	110	120	150	180	130	180	210	240
GEORGIA									
ATLANTA	90	110	110	150	180	120	170	210	240
SAVANNAH	100	120	130	160	190	140	190	220	250
HAWAII									
HILO	90	100	110	150	170	120	170	200	230
HONOLULU	110	110	150	180	190	170	220	250	270
IDAHO									
BOISE	120	130	140	190	210	160	210	260	290
POCATELLO	120	140	150	200	230	170	230	270	310
ILLINOIS									
CHICAGO	90	120	110	150	180	120	160	200	230
PEORIA	100	130	120	170	200	130	180	230	260
INDIANA									
INDIANAPOLIS	90	110	100	140	160	110	150	180	210
IOWA									
DES MOINES	110	130	120	170	200	130	190	230	260
KANSAS									
WICHITA	110	130	140	180	210	150	210	250	290
KENTUCKY									
LEXINGTON	90	110	100	140	170	110	150	190	220
LOUISVILLE	90	110	100	140	170	110	150	190	220
LOUISIANA									
NEW ORLEANS	90	110	120	150	180	130	170	210	240
SHREVEPORT	100	110	120	160	180	130	180	220	250
MAINE									
CARIBOU	90	120	100	140	170	110	150	190	230
PORTLAND	80	110	90	130	160	100	140	180	210
MARYLAND									
BALTIMORE	90	110	110	140	170	110	160	200	230
MASSACHUSETTS									
AMHERST	90	110	100	140	170	110	150	190	220
BOSTON	80	110	100	130	160	100	140	180	210
MICHIGAN									
LANSING	100	120	110	150	180	120	170	210	240
SAULT ST. MARIE	80	110	90	130	160	100	140	170	210

Number of Occupants	2		4			6 (or more)			
Collector Area (sq. ft.)	40	60	40	60	80	40	60	80	100
Location									
MINNESOTA									
MINN-ST. PAUL	100	120	110	150	190	120	170	210	240
MISSISSIPPI									
JACKSON	90	110	120	150	180	130	170	210	240
MISSOURI									
KANSAS CITY	100	120	120	160	190	130	180	220	260
ST. LOUIS	100	120	120	160	190	130	170	210	250
MONTANA									
BILLINGS	110	130	130	180	210	140	200	240	280
GREAT FALLS	110	130	130	170	200	140	190	230	270
NEBRASKA									
LINCOLN	110	140	130	180	210	140	200	250	280
NEVADA									
LAS VEGAS	120	130	170	210	220	190	250	290	300
RENO	140	150	180	230	250	200	260	320	350
NEW HAMPSHIRE									
CONCORD	80	110	90	130	160	100	140	170	210
NEW JERSEY									
ATLANTIC CITY	110	130	130	170	200	140	190	230	270
NEW MEXICO									
ALBUQUERQUE	140	140	180	220	240	190	260	310	340
NEW YORK									
ALBANY	100	120	110	150	180	120	160	200	240
NEW YORK	80	100	90	130	150	100	140	170	200
ROCHESTER	70	100	90	120	140	90	130	160	190
SYRACUSE	70	100	80	120	140	90	130	160	190
NORTH CAROLINA									
CAPE HATTERAS	100	120	120	160	180	130	170	210	240
RALEIGH	90	110	110	150	180	120	170	200	230
NORTH DAKOTA									
BISMARCK	110	140	130	180	210	140	190	240	280
OHIO									
CLEVELAND	80	100	90	120	150	90	130	170	190
COLUMBUS	80	100	90	130	160	100	140	170	200
OKLAHOMA									
OKLAHOMA CITY	110	120	130	170	200	140	190	230	270
TULSA	100	120	120	160	190	130	180	220	250
OREGON									
MEDFORD	100	110	120	160	180	130	180	210	240
PORTLAND	80	90	90	120	140	90	130	160	190
PENNSYLVANIA									
PHILADELPHIA	90	110	100	140	170	110	150	190	220
PITTSBURGH	100	120	120	160	190	130	180	220	250
STATE COLLEGE	90	120	110	150	180	110	160	200	230
RHODE ISLAND									
PROVIDENCE	90	110	100	130	160	100	150	180	210
SOUTH CAROLINA									
CHARLESTON	90	110	110	150	170	120	160	200	230
SOUTH DAKOTA									
RAPID CITY	110	140	140	180	210	140	200	250	280
TENNESSEE									
NASHVILLE	90	110	110	140	170	110	160	190	220
TEXAS									
AMARILLO	120	140	160	200	230	170	230	280	310
DALLAS	100	110	120	160	180	130	180	220	250
EL PASO	130	130	170	210	220	190	250	290	310
HOUSTON	90	100	110	140	160	110	160	190	220
UTAH									
SALT LAKE CITY	120	140	160	200	230	170	230	280	310
VERMONT									
BURLINGTON	90	120	110	140	170	110	160	200	230
VIRGINIA									
MT. WEATHER	100	130	120	160	200	130	180	220	260
NORFOLK	100	120	120	150	180	120	170	210	240
RICHMOND	90	110	110	150	170	120	160	200	230
WASHINGTON									
SEATTLE	80	100	90	130	150	100	140	170	200
SPOKANE	100	110	110	150	180	120	170	210	240
WEST VIRGINIA									
CHARLESTON	80	100	90	130	150	100	140	170	200
WISCONSIN									
MADISON	100	120	110	150	180	120	170	210	240
WYOMING									
CASPER	130	150	170	220	250	180	250	300	340
CHEYENNE	130	150	160	210	240	170	230	280	330

TABLE B. Cumulative Savings Factors

Note: To use this table, you must choose a Fuel Price Escalation Rate that seems reasonable to you. Economic predictions are frequently based on the assumption that fuel prices will rise about 10% annually in the years ahead, but published estimates of the escalation rate vary from 4% to 20%.

Time Period (years)	Fuel Price Escalation Rate (% per year)								
	4	6	8	10	12	14	16	18	20
1	1.0	1.0	1.0	1.0	1.0	1.0	1.0	1.0	1.0
5	5.4	5.6	5.9	6.1	6.4	6.6	6.9	7.2	7.4
10	12.0	13.2	14.5	15.9	17.5	19.3	21.3	23.5	26.0
15	20.0	23.3	27.2	31.8	37.3	43.8	51.7	61.0	72.0
20	29.8	36.8	45.8	57.3	72.1	91.0	115.4	146.6	186.7

TABLE C. Suggested Collector Sizes
For Solar Water Heating Systems

Note: The collector area indicated in each case can be expected to provide, at the lowest possible cost per unit of heat produced, at least 50 percent of the hot water used annually by the average household of that size. A collector panel typically has 20 square feet of collecting surface.

Location	Number of Users 2	4	6	Location	Number of Users 2	4	6
	Square Feet				Square Feet		
BIRMINGHAM, AL	40	60	80	BILLINGS, MT	40	60	80
FAIRBANKS, AK	60	80	100	GREAT FALLS, MT	40	60	80
TUCSON, AZ	40	60	60	LINCOLN, NE	40	60	80
LITTLE ROCK, AR	40	60	80	LAS VEGAS, NV	40	60	60
LOS ANGELES, CA	40	60	80	RENO, NV	40	60	80
SACRAMENTO, CA	40	60	60	CONCORD, NH	60	80	100
SAN FRANCISCO, CA	40	60	80	ATLANTIC CITY, NJ	40	60	80
DENVER, CO	40	60	80	ALBUQUERQUE, NM	40	60	60
GRAND JUNCTION, CO	40	60	80	ALBANY, NY	60	80	100
HARTFORD, CT	60	80	100	NEW YORK, NY	60	80	100
WILMINGTON, DE	60	80	100	ROCHESTER, NY	60	80	100
WASHINGTON, DC	60	80	100	SYRACUSE, NY	60	80	100
JACKSONVILLE, FL	40	60	80	CAPE HATTERAS, NC	40	60	80
MIAMI, FL	40	60	80	RALEIGH, NC	40	60	80
TALLAHASSEE, FL	40	60	80	BISMARCK, ND	40	80	100
TAMPA, FL	40	60	80	CLEVELAND, OH	60	80	100
ATLANTA, GA	40	60	80	COLUMBUS, OH	60	80	100
SAVANNAH, GA	40	60	80	OKLAHOMA CITY, OK	40	60	80
HILO, HI	40	60	80	TULSA, OK	40	60	80
HONOLULU, HI	40	60	60	MEDFORD, OR	40	60	80
BOISE, ID	40	60	80	PORTLAND, OR	60	80	100
POCATELLO, ID	40	60	80	PHILADELPHIA, PA	60	80	100
CHICAGO, IL	60	80	100	PITTSBURGH, PA	40	80	80
PEORIA, IL	40	60	80	STATE COLLEGE, PA	60	80	100
INDIANAPOLIS, IN	60	80	100	PROVIDENCE, RI	60	80	100
DES MOINES, IA	40	80	80	CHARLESTON, SC	40	60	80
WICHITA, KS	40	60	80	RAPID CITY, SD	40	60	80
LEXINGTON, KY	60	80	100	NASHVILLE, TN	40	60	80
LOUISVILLE, KY	60	80	100	AMARILLO, TX	40	60	80
NEW ORLEANS, LA	40	60	80	DALLAS, TX	40	60	80
SHREVEPORT, LA	40	60	80	EL PASO, TX	40	60	60
CARIBOU, ME	60	80	100	HOUSTON, TX	40	60	80
PORTLAND, ME	60	80	100	SALT LAKE CITY, UT	40	60	80
BALTIMORE, MD	60	80	100	BURLINGTON, VT	60	80	100
AMHERST, MA	60	80	100	MT. WEATHER, VA	60	80	80
BOSTON, MA	60	80	100	NORFOLK, VA	40	60	80
LANSING, MI	60	80	100	RICHMOND, VA	60	80	80
SAULT STE. MARIE, MI	60	80	100	SEATTLE, WA	60	80	100
MINN.-ST. PAUL, MN	60	80	100	SPOKANE, WA	40	60	80
JACKSON, MS	40	60	80	CHARLESTON, WV	60	80	100
KANSAS CITY, MO	40	60	80	MADISON, WI	60	80	100
ST. LOUIS, MO	40	60	80	CASPER, WY	40	60	80
				CHEYENNE, WY	40	60	80

TABLE D. Typical Costs In 1979 For Solar Hot Water Systems

Note: This table, which is based on information from the HUD solar demonstration program and from industry sources, lists approximate installed costs, *before* tax credits, that were typical in 1979 in areas where freeze protection is necessary. Somewhat lower costs may be expected in the warm southern states. If you need help in estimating the collector size best suited to your hot water use, refer to Table C. Areas given below are multiples of 20 square feet, the average size of a collector panel.

Collector Size (Sq. Ft.)	Installed Cost For New Homes $	Installed Cost For Existing Homes $
20	1450	1600
40	1900	2050
60	2300	2550
80	2750	3000
100	3200	3500

TABLE E. Federal Energy Tax Credits

Note: To find the credit on values not appearing in the table, follow this simple method:

If the installed cost is $10,000 or less, multiply it by 40 percent.

(For a copy of I.R.S. Publication 903, "Energy Credits for Individuals," return the request form at the back of the guide.)

If Your Installed Solar Hot Water System Cost Is: $	Then Your Tax Credit Is: $
1000	400
1250	500
1450	580
1600	640
1750	700
1900	760
2050	820
2300	920
2550	1020
2750	1100
3000	1200
3200	1280
3500	1400

TABLE F. Annual Loan Payment Factors For Mortgage Loans

Term of Loan (Years)	Interest Rate (%)								
	9	9½	10	10½	11	11½	12	13	14
20	.108	.112	.116	.120	.124	.128	.132	.141	.149
25	.101	.105	.109	.113	.118	.122	.126	.135	.145
30	.097	.101	.105	.110	.114	.119	.123	.133	.142

TABLE G. Year In Which Favorable Cash Flow Begins

Note: Use the same Fuel Price Escalation Rate that you chose when referring to Table B in Step 1.

Loan Cost/ Savings Ratio	Fuel Price Escalation Rate (% per year)								
	4	6	8	10	12	14	16	18	20
1.0*	1	1	1	1	1	1	1	1	1
1.1	4	3	3	2	2	2	2	2	2
1.2	6	5	4	3	3	3	3	3	2
1.3	8	6	5	4	4	4	3	3	3
1.4	10	7	6	5	4	4	4	4	3
1.5	12	8	7	6	5	5	4	4	4
1.6	13	10	8	6	6	5	5	4	4
1.7	15	11	8	7	6	6	5	5	5
1.8	16	12	9	8	7	6	5	5	5
1.9	18	13	10	8	7	6	6	5	5
2.0	19	13	11	9	8	7	6	6	5
2.1	20	14	11	9	8	7	6	6	6
2.2	22	15	12	10	8	8	7	6	6
2.3	23	16	12	10	9	8	7	7	6
2.4	24	17	13	11	9	8	7	7	6
2.5	25	17	13	11	10	8	8	7	7
2.6	26	18	14	12	10	9	8	7	7
2.7	27	19	14	12	10	9	8	8	7
2.8	28	19	15	12	11	9	8	8	7
2.9	29	20	15	13	11	10	9	8	7
3.0**	30	20	16	13	11	10	9	8	8

*If the ratio is less than 1.0, favorable cash flow begins immediately.
**If the ratio is greater than 3.0, the system is unlikely to be economical under any conditions.

TABLE H. Year In Which Accumulated Savings Exceed Total Costs

Cost/Savings Ratio	Fuel Escalation Rate (% per year)								
	4	**6**	**8**	**10**	**12**	**14**	**16**	**18**	**20**
1.	1	1	1	1	1	1	1	1	1
2.	2	2	2	2	2	2	2	2	2
3.	3	3	3	3	3	3	3	3	3
4.	4	4	4	4	4	4	4	4	4
5.	5	5	5	5	5	5	4	4	4
6.	6	6	6	5	5	5	5	5	5
7.	7	7	6	6	6	6	6	5	5
8.	8	7	7	7	6	6	6	6	6
9.	8	8	8	7	7	7	7	6	6
10.	9	9	8	8	7	7	7	7	7
15.	12	12	11	10	10	9	9	8	8
20.	15	14	13	12	11	11	10	10	9
25.	18	16	15	14	13	12	11	11	10
30.	21	18	16	15	14	13	12	12	11
35.	23	20	18	16	15	14	13	13	12
40.	25	11	19	17	16	15	14	13	13
45.	27	23	20	18	17	16	15	14	13
50.	29	24	21	19	18	16	15	14	14
55.	30	26	22	20	18	17	16	15	14
60.	**	27	23	21	19	18	16	15	15

Note: Use the same Fuel Price Escalation Rate that you chose when referring to Table B in Step 1.

**More than 30 years.

TABLE J. Short-Term Loan Payment Factors

Term of Loan (Years)	Interest Rate (%)							
	8	9	10	11	12	13	14	15
1	1.04	1.05	1.06	1.06	1.07	1.07	1.08	1.08
2	1.09	1.10	1.11	1.12	1.13	1.14	1.15	1.16
3	1.14	1.14	1.16	1.18	1.20	1.21	1.23	1.25
4	1.17	1.19	1.22	1.24	1.26	1.29	1.31	1.34
5	1.22	1.25	1.28	1.31	1.34	1.37	1.40	1.43
6	1.26	1.30	1.33	1.37	1.41	1.45	1.48	1.52
7	1.31	1.35	1.39	1.44	1.48	1.53	1.57	1.62
8	1.36	1.41	1.46	1.51	1.56	1.61	1.67	1.72
9	1.41	1.46	1.52	1.58	1.64	1.70	1.76	1.83
10	1.46	1.52	1.59	1.65	1.72	1.79	1.86	1.94
11	1.51	1.58	1.65	1.73	1.81	1.88	1.96	2.05
12	1.56	1.64	1.72	1.80	1.89	1.98	2.07	2.16
13	1.61	1.70	1.79	1.88	1.98	2.08	2.18	2.28
14	1.67	1.76	1.86	1.96	2.07	2.18	2.29	2.40
15	1.72	1.83	1.94	2.05	2.16	2.28	2.40	2.52

TABLE K. Collector Performance Adjustment Factors

Note: Annual Energy Savings figures in Table A are based on *averaged* collector performance data. To adjust for *specific* performance data: 1) In Table A, find the number of units of electricity, natural gas, or oil that a typical solar water heater would save you annually. 2) Consult the charted curve representing a specific collector's thermal performance when tested according to ASHRAE Standard 93-77. Note the collector efficiency value at the point where the performance curve intercepts the vertical axis (see example below). 3) In this table, find the adjustment factor corresponding to that efficiency value and to the size of your household and proposed collector area. 4) Multiply the figure selected from Table A by this adjustment factor to find the estimated number of kilowatt-hours, therms, or gallons that a specific collector would save annually. Enter this number on line 1a, Step 1, of the worksheet.

Persons in Household	2 Persons		4 Persons			6 Persons		
Collector Size	40 Sq Ft	60 Sq Ft	40 Sq Ft	60 Sq Ft	80 Sq Ft	60 Sq Ft	80 Sq Ft	100 Sq Ft
Efficiency Value								
0.65	0.90	0.92	0.88	0.89	0.90	0.88	0.89	0.90
0.67	0.92	0.94	0.91	0.92	0.92	0.91	0.92	0.92
0.69	0.95	0.96	0.94	0.95	0.95	0.94	0.95	0.95
0.71	0.98	0.98	0.97	0.97	0.97	0.97	0.97	0.98
0.73	1.00	1.00	1.00	1.00	1.00	1.00	1.00	1.00
0.75	1.02	1.01	1.03	1.02	1.02	1.03	1.03	1.02
0.77	1.04	1.03	1.06	1.05	1.04	1.06	1.05	1.04
0.79	1.06	1.04	1.09	1.07	1.06	1.08	1.08	1.06
0.81	1.09	1.06	1.11	1.10	1.09	1.11	1.10	1.09
0.83	1.11	1.07	1.14	1.12	1.11	1.14	1.13	1.11
0.85	1.13	1.08	1.17	1.15	1.13	1.17	1.15	1.13

Example

The Wainwrights of Los Angeles, a family of six, are comparing the economics of two packaged solar water heating systems which differ in collector efficiency. Their dealer has supplied test data for both types of collectors, combining the two performance curves on a single chart (below). Following the procedure described in TABLE K, the Wainwrights find that:

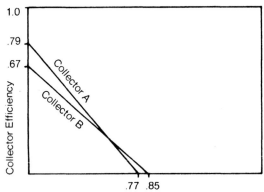

Ratio of Temperature Difference to Insolation
$$\left[\frac{T - Ta}{I} \right]$$

1) A typical solar water heater would save them 180 therms of natural gas, per TABLE A.

2) At the point where its performance curve intercepts the vertical axis, Collector A has an efficiency value of .79; Collector B has a value of .67.

3) For a six-person family planning 80 square feet of collector, an adjustment factor of 1.08 corresponds to Collector A's efficiency value; in the case of Collector B, the factor is .92.

4) Collector A could save about 195 therms annually (180 x 1.08 = 194.4), and Collector B about 165 therms (180 x .92 = 165.6). To estimate costs and savings from the packaged system using Collector A, enter 195 therms on line 1a, STEP 1, of the worksheet. For the system using Collector B, substitute 165 therms on line 1a.

Printed in the United States
38782LVS00002B/425